HEINEMANN MODULAR MATHEMATICS
for
EDEXCEL AS AND A-LEVEL

Revise for Core Mathematics 1

Greg Attwood Alistair Macpherson Bronwen Moran
Joe Petran Keith Pledger Geoff Staley Dave Wilkins

Heinemann
Inspiring generations

Heinemann Educational Publishers
Halley Court, Jordan Hill, Oxford OX2 8EJ
Part of Harcourt Education

Heinemann is the registered trademark of
Harcourt Education Limited

First published 2005

09 08 07 06
10 9 8 7 6 5 4 3

British Library Cataloguing in Publication Data is available
from the British Library on request.

10-digit ISBN: 0 435511 22 X
13-digit ISBN: 978 0 435511 22 7

Designed by Bridge Creative Services
Typeset by Tech-set Ltd

Illustrated by Tech-Set Ltd

Cover design by Bridge Creative Services

Printed in China by CTPS.

Acknowledgements
Every effort has been made to contact copyright holders of material reproduced in this book. Any omissions will be rectified in subsequent printings if notice is given to the publishers.

About this book

This book is designed to help you get your best possible grade in your Core 1 examination. The authors are Chief and Principal examiners, and have a good understanding of Edexcel's requirements.

Revise for Core 1 covers the key topics that are tested in the Core 1 exam paper. You can use this book to help you revise at the end of your course, or you can use it throughout your course alongside the course textbook, *Heinemann Modular Mathematics for Edexcel AS and A-level Core 1*, which provides complete coverage of the syllabus.

Helping you prepare for your exam

To help you prepare, each topic offers you:

- **Key points to remember** – summarise the mathematical ideas you need to know and be able to use.
- **Worked examples and examination questions** – help you understand and remember important methods, and show you how to set out your answers clearly.
- **Revision exercises** – help you practise using these important methods to solve problems. Exam-level questions are included so you can be sure you are reaching the right standard, and answers are given at the back of the book so you can assess your progress.
- **Test Yourself questions** – help you see where you need extra revision and practice. If you do need extra help, they show you where to look in the *Heinemann Modular Mathematics for Edexcel AS and A-level Core 1* textbook and which example to refer to in this book.

Exam practice and advice on revising

Examination style paper – this paper at the end of the book provides a set of questions of examination standard. It gives you an opportunity to practise taking a complete exam before you meet the real thing. The answers are given at the back of the book.

How to revise – for advice on revising before the exam, read the How to revise section on the next page.

How to revise using this book

Making the best use of your revision time

The topics in this book have been arranged in a logical sequence so you can work your way through them from beginning to end. But **how** you work on them depends on how much time there is between now and your examination.

If you have plenty of time before the exam then you can **work through each topic in turn**, covering the key points and worked examples before doing the revision exercises and test yourself questions.

If you are short of time then you can **work through the Test Yourself sections** first, to help you see which topics you need to do further work on.

However much time you have to revise, make sure you break your revision into short blocks of about 40 minutes, separated by five- or ten-minute breaks. Nobody can study effectively for hours without a break.

Using the Test Yourself sections

Each Test Yourself section provides a set of key questions. Try each question:

- If you can do it and get the correct answer, then move on to the next topic. Come back to this topic later to consolidate your knowledge and understanding by working through the key points, worked examples and revision exercises.

- If you cannot do the question, or get an incorrect answer or part answer, then work through the key points, worked examples and revision exercises before trying the Test Yourself questions again. If you need more help, the cross-references beside each Test Yourself question show you where to find relevant information in the *Heinemann Modular Mathematics for Edexcel AS and A-level Core 1* textbook and which example in *Revise for C1* to refer to.

Reviewing the key points

Most of the key points are straightforward ideas that you can learn: try to understand each one. Imagine explaining each idea to a friend in your own words, and say it out loud as you do so. This is a better way of making the ideas stick than just reading them silently from the page.

As you work through the book, remember to go back over key points from earlier topics at least once a week. This will help you to remember them in the exam.

Algebra and functions

1

Key points to remember

1 You can simplify expressions by collecting like terms.

2 You can simplify expressions by using the rules of indices.

(i) $a^m \times a^n = a^{m+n}$

(ii) $a^m \div a^n = a^{m-n}$

(iii) $a^{-m} = \dfrac{1}{a^m}$

(iv) $a^{\frac{1}{m}} = \sqrt[m]{a}$

(v) $a^{\frac{n}{m}} = \sqrt[m]{a^n}$

(vi) $(a^m)^n = a^{mn}$

(vii) $a^0 = 1$

3 You can expand an expression by multiplying each term inside the bracket by the terms outside the bracket.

4 Factorising expressions is the opposite of expanding expressions.

5 A quadratic expression has the form $ax^2 + bx + c$, where a, b and c are constants and $a \neq 0$.

6 $x^2 - y^2 = (x + y)(x - y)$. This is called a **difference of squares**.

7 You can write a number exactly using surds.

8 The square root of a prime number is a surd.

9 You can manipulate surds using these rules:

(i) $\sqrt{ab} = \sqrt{a} \times \sqrt{b}$ (ii) $\sqrt{\dfrac{a}{b}} = \dfrac{\sqrt{a}}{\sqrt{b}}$.

10 The rules to rationalise surds are:

(i) for fractions in the form $\dfrac{1}{\sqrt{a}}$, multiply the top and bottom by $\sqrt{a}$

(ii) for fractions in the form $\dfrac{1}{a + \sqrt{b}}$, multiply the top and bottom by $a - \sqrt{b}$

(iii) for fractions in the form $\dfrac{1}{a - \sqrt{b}}$, multiply the top and bottom by $a + \sqrt{b}$.

Example 1

Simplify

(a) $2x^3 \times 3x^4 \times 5x^7$ **(b)** $(2y^4)^3 \div 4y^2$

(a) $2x^3 \times 3x^4 \times 5x^7 = 2 \times 3 \times 5 \times x^3 \times x^4 \times x^7$

$= 30 \times x^{3+4+7}$

$= 30x^{14}$

(b) $(2y^4)^3 \div 4y^2 = 2^3 \times y^{4\times 3} \div 4y^2$

$= \frac{8}{4} \times y^{4\times 3} \div y^2$

$= 2 \times y^{12-2}$

$= 2y^{10}$

Rewrite the expression with the numbers together and the x-terms together.

Using **2** (i): $x^3 \times x^4 \times x^7 = x^{3+4+7}$

Using **2** (vi)

$2^3 = 8$

Using **2** (ii): $y^{12} \div y^2 = y^{12-2}$

Example 2

Factorise

(a) $6x^2 - x - 15$ **(b)** $4x^2 - 9$

(a) $6x^2 - x - 15$

$ac = -90$

$-10 + 9 = -1$

So $6x^2 - 10x + 9x - 15$

$= 2x(3x - 5) + 3(3x - 5)$

$= (3x - 5)(2x + 3)$

(b) $4x^2 - 9$

$= (2x)^2 - 3^2$

$= (2x + 3)(2x - 3)$

This is a quadratic, where $a = 6$, $b = -1$ and $c = -15$. You need to find two brackets that multiply together to give $6x^2 - x - 15$.

To do this work out ac

Work out the two factors of ac that add to give you b

Factorise using **4**

Factorise by $(3x - 5)$

Using **6**

Example 3

Evaluate

(a) $27^{\frac{2}{3}}$ **(b)** $\left(\frac{81}{16}\right)^{-\frac{3}{4}}$

(a) $27^{\frac{2}{3}} = (\sqrt[3]{27})^2$

$= (3)^2$

$= 9$

Using **2** (v)

$3 \times 3 \times 3 = 27$ So $\sqrt[3]{27} = 3$

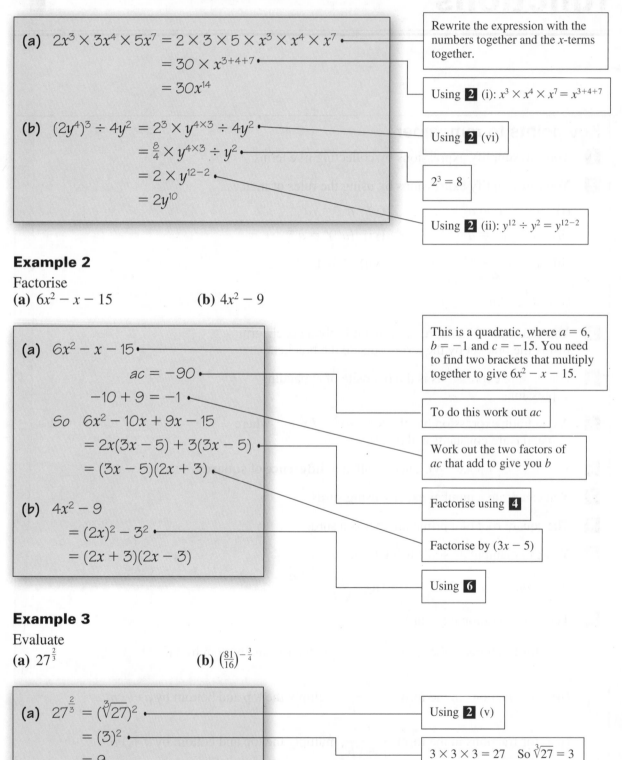

(b) $\left(\frac{81}{16}\right)^{-\frac{3}{4}} = \left(\frac{16}{81}\right)^{\frac{3}{4}}$ — Using **2** (iii)

$= \frac{(\sqrt[4]{16})^3}{(\sqrt[4]{81})^3}$ — Using **2** (v)

$= \frac{2^3}{3^3}$ — As $2 \times 2 \times 2 \times 2 = 16$, $\sqrt[4]{16} = 2$ and $3 \times 3 \times 3 \times 3 = 81$, $\sqrt[4]{81} = 3$

$= \frac{8}{27}$

Example 4

Simplify

(a) $\sqrt{40}$ **(b)** $\sqrt{40} + \sqrt{90}$

(a) $\sqrt{40} = \sqrt{4 \times 10}$ — Using **9**

$= 2\sqrt{10}$ — $\sqrt{4} = 2$

(b) $\sqrt{40} + \sqrt{90} = 2\sqrt{10} + \sqrt{9} \times \sqrt{10}$ — Using **9**

$= 2\sqrt{10} + 3\sqrt{10}$

$= 5\sqrt{10}$

Worked exam style question 1

(a) Express 80 in the form $a\sqrt{5}$, where a is an integer.

(b) Express $(4 - \sqrt{5})^2$ in the form $b + c\sqrt{5}$, where b and c are integers.

(a) $\sqrt{80} = \sqrt{16 \times 5}$

$= 4\sqrt{5}$ — Using **9**; $4 \times 4 = 16$

(b) $(4 - \sqrt{5})^2 = (4 - \sqrt{5})(4 - \sqrt{5})$

$= 16 - 4\sqrt{5} - 4\sqrt{5} + (\sqrt{5})^2$

$= 16 - 8\sqrt{5} + (\sqrt{5})^2$

$= 16 - 8\sqrt{5} + 5$ — $(\sqrt{5})^2 = 5$

$= 21 - 8\sqrt{5}$

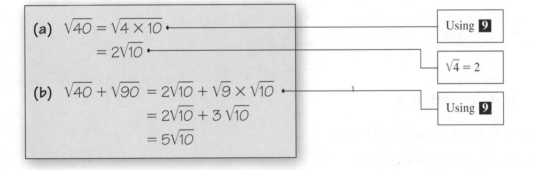

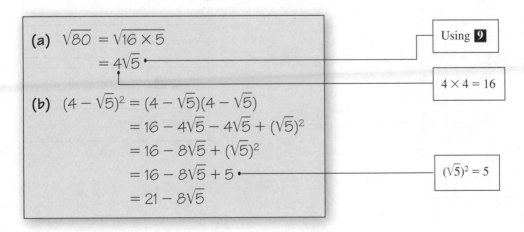

Worked exam style question 2

Rationalise the denominator for $\dfrac{\sqrt{3}-\sqrt{5}}{\sqrt{5}-\sqrt{3}}$

$$\frac{(\sqrt{3}-\sqrt{5})(\sqrt{5}+\sqrt{3})}{(\sqrt{5}-\sqrt{3})(\sqrt{5}+\sqrt{3})}$$

Using **10**: multiply top and bottom by $(\sqrt{5}+\sqrt{3})$

$$=\frac{\sqrt{3}\times\sqrt{5}+\sqrt{3}\sqrt{3}-\sqrt{5}\sqrt{5}-\sqrt{5}\sqrt{3}}{\sqrt{5}\sqrt{5}+\sqrt{5}\sqrt{3}-\sqrt{3}\sqrt{5}-\sqrt{3}\sqrt{3}}$$

Using **9**

$$=\frac{\sqrt{15}+\sqrt{9}-\sqrt{25}-\sqrt{15}}{\sqrt{25}+\sqrt{15}-\sqrt{15}-\sqrt{9}}$$

You should be able to miss this step out and write this straight down

$$=\frac{-5+3}{5-3}$$

$$=\frac{-2}{2}$$

$$=-1$$

Worked exam style question 3

(a) Given that $27 = 3^k$, write down the value of k.

(b) Given that $9^y = 27^{2y-1}$, find the value of y.

(a) $27 = 3^k$

$3^3 = 3^k$

$k = 3$ — Comparing powers

(b) $9^y = 27^{2y-1}$

$(3^2)^y = (3^3)^{2y-1}$ — $9 = 3^2$, $27 = 3^3$

$3^{2y} = 3^{6y-3}$ — Using **2** (vi)

$2y = 6y - 3$ — Comparing powers

$4y = 3$

So $y = \frac{3}{4}$

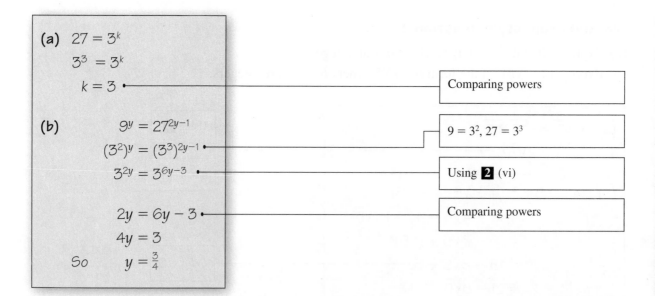

Revision exercise 1

1 Simplify:

(a) $3x^2 \times 4x^5 \times 2x^3$ **(b)** $(8y^4)^2 \div 4y^3$

2 Factorise:

(a) $9x^2 - 4y^2$ **(b)** $6y^2 - 15y$

(c) $12x^2 - 14x - 6$ **(d)** $2x^2 + 5x + 3$

(e) $6 + x - 2x^2$

Hint: take 2 outside as a common factor

3 Evaluate:

(a) $16^{\frac{1}{2}}$ **(b)** $64^{\frac{2}{3}}$ **(c)** $(343)^{-\frac{2}{3}}$

(d) $\left(\frac{4}{25}\right)^{-\frac{1}{2}}$ **(e)** $\left(\frac{8}{27}\right)^{\frac{2}{3}}$ **(e)** $\left(\frac{512}{27}\right)^{-\frac{2}{3}}$

4 Evaluate $\sqrt{245} - 3\sqrt{45} + 2\sqrt{20}$, giving your answer in terms of $a\sqrt{5}$ where a is a constant.

5 Simplify:

(a) $\sqrt{2} \times \sqrt{8}$ **(b)** $\sqrt{6} \times \sqrt{8} \times \sqrt{12}$

6 Rationalise the denominators of:

(a) $\dfrac{1}{3 - \sqrt{5}}$ **(b)** $\dfrac{\sqrt{2}}{\sqrt{6} - \sqrt{2}}$ **(c)** $\dfrac{3\sqrt{5} + 2}{2\sqrt{5} - 4}$

7 Simplify $\dfrac{1}{\sqrt{3} + 1} + \dfrac{1}{\sqrt{3} - 1}$

8 **(a)** Express $\sqrt{112}$ in the form $a\sqrt{7}$, where $\sqrt{a}$ is an integer.

(b) Express $(3 - \sqrt{5})^2$ in the form $b + c\sqrt{5}$, where b and c are integers.

9 **(a)** Given that $8 = 2^k$, write down the value of k.

(b) Given that $4^x = 8^{x-1}$, find the values of x. *E*

10 Find the value of:

(a) $81^{\frac{1}{2}}$ **(b)** $81^{\frac{3}{4}}$ **(c)** $81^{-\frac{3}{4}}$

E

Test yourself	What to review
	If your answer is incorrect
1 Evaluate $\left(\frac{27}{8}\right)^{\frac{2}{3}}$	*Review Heinemann Book C1 pages 10–11* *Revise for C1 page 2 Example 3*
2 Simplify the expressions **(a)** $3(x + 4y^2) - 2(3x + y^2)$ **(b)** $4x^2 \times 3x^5 \div 6x^3$	*(a) Review Heinemann Book C1 page 1* *(b) Review Heinemann Book C1 pages 2–3* *Revise for C1 page 2 Example 1*
3 Expand $5x^2(3x - 2) - 3x^2(2x - 5)$	*Review Heinemann Book C1 pages 3–4*
4 Factorise completely **(a)** $4x^2 + 10x$ **(b)** $16x^2 - 9y^2$ **(c)** $6x^2 - 7x - 5$	*(a) Review Heinemann Book C1 page 4* *(b) Review Heinemann Book C1 pages 4–6* *Revise for C1 page 2 Example 2b* *(c) Review Heinemann Book C1 pages 5–6* *Revise for C1 page 2 Example 2a*
5 Simplify **(a)** $\sqrt{72}$ **(b)** $2\sqrt{12} + \sqrt{48} + 3\sqrt{75}$	*Review Heinemann Book C1 pages 9–10* *Revise for C1 page 3 Example 4*
6 Rationalise the denominators of **(a)** $\frac{4}{1 - \sqrt{3}}$ **(b)** $\frac{\sqrt{5} + 2}{\sqrt{7} - 3}$	*Review Heinemann Book C1 pages 10–11* *Revise for C1 page 4* *Worked exam style question 2*
7 Express $(3 - \sqrt{11})^2$ in the form $a + b\sqrt{11}$	*Review Heinemann Book C1 pages 10–11* *Revise for C1 page 3* *Worked exam style question 1*

Test yourself answers

1 $\frac{9}{4}$ **2 (a)** $10y^2 - 3x$ **(b)** $2x^4$ **3** $9x^3 + 5x^2$ **4 (a)** $2x(2x + 5)$ **(b)** $(4x + 3y)(4x - 3y)$ **(c)** $(2x + 1)(3x - 5)$

5 (a) $6\sqrt{2}$ **(b)** $23\sqrt{3}$ **6 (a)** $-2(1 + \sqrt{3})$ **(b)** $\frac{1}{2}(6 + \sqrt{35} + 2\sqrt{7} + 3\sqrt{5})$ **7** $20 - 6\sqrt{11}$

Quadratic functions 2

Key points to remember

1 The general form of a quadratic equation is $y = ax^2 + bx + \mathrm{c}$ where a, b, c are constants and $a \neq 0$.

2 Quadratic equations can be solved by:

(i) factorisation

(ii) completing the square:

$$x^2 + bx = \left(x + \frac{b}{2}\right)^2 - \left(\frac{b}{2}\right)^2$$

(iii) using the formula

$$\frac{x = -b \pm \sqrt{(b^2 - 4ac)}}{2a}$$

3 A quadratic equation has two solutions, which may be equal.

4 To sketch a quadratic graph:

(i) decide on the shape

$a > 0$ ∪

$a < 0$ ∩

(ii) work out the x-axis and y-axis crossing points

(iii) check the general shape by considering the discriminant $b^2 - 4ac$.

5 A quadratic equation has:

(i) equal roots when $b^2 = 4ac$

(ii) real roots when $b^2 \geqslant 4ac$

(iii) real different roots when $b^2 > 4ac$

(iv) no real roots when $b^2 < 4ac$.

Example 1

Solve the following equations by factorisation
(a) $x^2 + 5x - 6 = 0$ **(b)** $6x^2 + 7x - 3 = 0$ **(c)** $x^3 + 3x^2 + 2x = 0$

(a) $x^2 + 5x - 6 = 0$

$(x + 6)(x - 1) = 0$ — Using **2**: factorise to solve the equation

So either $x + 6 = 0 \Rightarrow x = -6$ — Either part of the product could be zero

or $x - 1 = 0 \Rightarrow x = 1$

So $x = -6$ or $x = 1$ — Using **3**

(b) $6x^2 + 7x - 3 = 0$ — Using **2**: factorise to solve the equation

$(3x - 1)(2x + 3) = 0$

So either $3x - 1 = 0 \Rightarrow x = \frac{1}{3}$

or $2x + 3 = 0 \Rightarrow x = -\frac{3}{2}$

So $x = \frac{1}{3}$ or $x = -\frac{3}{2}$

(c) $x^3 + 3x^2 + 2x = 0$

$x(x^2 + 3x + 2) = 0$ — Factorise by x

$x(x + 2)(x + 1) = 0$ — Using **2**: factorise $x^2 + 3x + 2$ to solve the equation

So $x = 0$ or $x = -2$ or $x = 1$

Example 2

Complete the square for **(a)** $x^2 + 5x$ **(b** $3x^2 + 7x + 5$

(a) $x^2 + 5x$ — Using **2**: complete the square

$= \left(x + \frac{5}{2}\right)^2 - \left(\frac{5}{2}\right)^2$

$= \left(x + \frac{5}{2}\right)^2 - \frac{25}{4}$

(b) $3x^2 + 7x - 5$

$= 3\left(x^2 + \frac{7}{3}x - \frac{5}{3}\right)$ — Taking out a factor of 3

$= 3\left[\left(x + \frac{7}{6}\right)^2 - \left(\frac{7}{6}\right)^2 - \frac{5}{3}\right]$ — Using **2**: complete the square

$= 3\left[\left(x + \frac{7}{6}\right)^2 - \frac{49}{36} - \frac{60}{36}\right]$

$= 3\left[\left(x + \frac{7}{6}\right)^2 - \frac{109}{76}\right]$

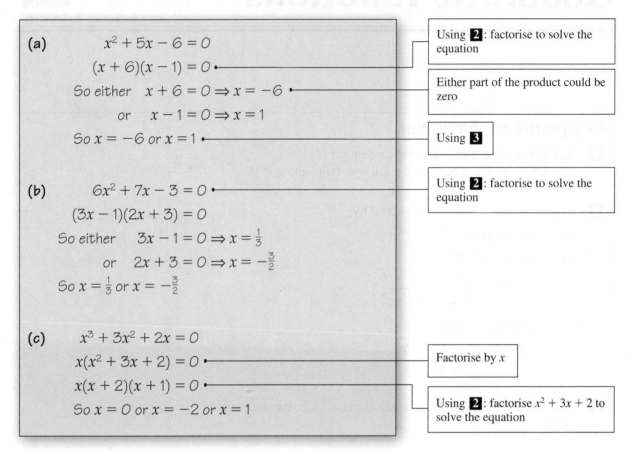

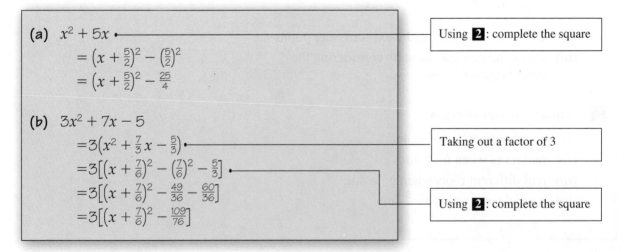

Example 3

Solve $2x^2 - 7x - 7 = 0$, leaving your answer in surd form.

$2x^2 - 7x + 4 = 0$

Using the formula $x = \dfrac{-b \pm \sqrt{b^2 - 4ac}}{2a}$

> Using **2**

$$x = \frac{-(-7) \pm \sqrt{(-7)^2 - 4(2)(-7)}}{2 \times 2}$$

> Using **1**:
> $a = 2$, $b = -7$ and $c = -7$

$$= \frac{+7 \pm \sqrt{49 + 56}}{4}$$

$$= \frac{+7 \pm \sqrt{105}}{4}$$

So $x = \dfrac{+7 + \sqrt{105}}{4}$ or $x = \dfrac{+7 - \sqrt{105}}{4}$

> Leave your answer in surd form because this is a non-calculator unit.

Worked exam style question 1

(a) By completing the square, find in terms of k, the roots of the equation

$x^2 + 4kx - 3 = 0$

(b) Show that, for all values of k, the roots of $x^2 + 4kx - 3 = 0$ are real and different.

(a) $x^2 + 4kx - 3 = 0$

$(x + 2k)^2 - (2k)^2 - 3 = 0$

> Using **2**: complete the square

$(x + 2k)^2 - 4k^2 - 3 = 0$

$(x + 2k)^2 = 4k^2 + 3$

$(x + 2k) = \sqrt{4k^2 + 3}$

$x = -2k \pm \sqrt{4k^2 + 3}$

(b) For real different roots $b^2 > 4ac$

Here $b^2 = 16k^2$

and $4ac = 4(1)(-3) = -12$

> Using **5**: where
> $a = 1$, $b = 4k$ and $c = -3$

So $16k^2 > 0 \quad \forall k \in \mathbb{R}$

and $4ac = -12$ which is < 0

therefore $b^2 > 4ac \quad \forall k \in \mathbb{R}$

Hence the roots are real and different.

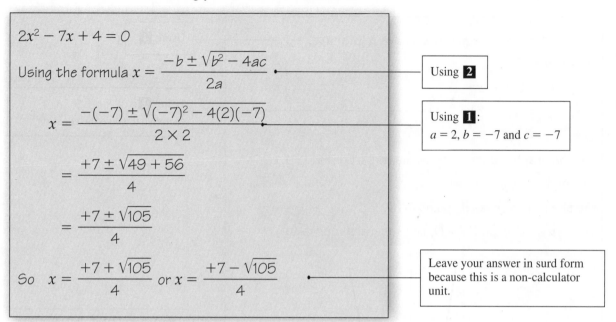

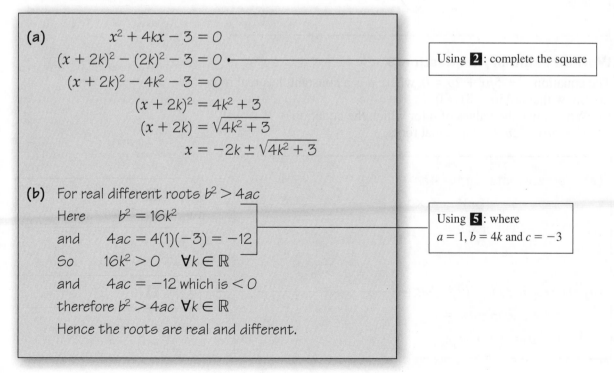

Worked exam style question 2

Sketch the graph of $y = 2x^2 - x - 3$

$a = 2$ so the graph will have a U shape.

When $y = 0$

$0 = 2x^2 - x - 3$

$0 = (2x - 3)(x + 1)$

So $x = \frac{3}{2}$ or $x = -1$

So the graph crosses the x-axis at $(\frac{3}{2}, 0)$ and $(-1, 0)$

When $x = 0, y = -3$

So the y-axis crossing point is $(0, -3)$

As $b^2 > 4ac$ and $a > 0$, the general shape is U.

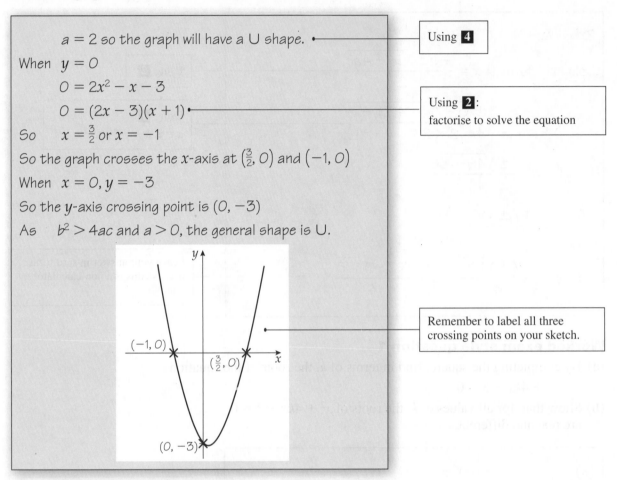

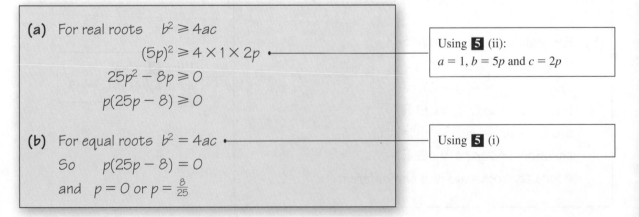

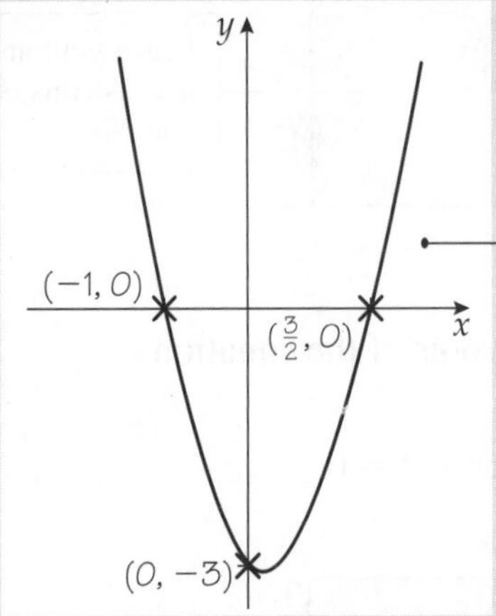

Using **4**

Using **2**: factorise to solve the equation

Remember to label all three crossing points on your sketch.

Worked exam style question 3

The equation $x^2 + 5px + 2p = 0$, where p is a constant, has real roots.

(a) Show that $p(25p - 8) \geqslant 0$

(b) Write down the values of p for which the equation $x^2 + 5px + 2p = 0$ has equal roots.

(a) For real roots $b^2 \geqslant 4ac$

$(5p)^2 \geqslant 4 \times 1 \times 2p$

$25p^2 - 8p \geqslant 0$

$p(25p - 8) \geqslant 0$

(b) For equal roots $b^2 = 4ac$

So $p(25p - 8) = 0$

and $p = 0$ or $p = \frac{8}{25}$

Using **5** (ii): $a = 1$, $b = 5p$ and $c = 2p$

Using **5** (i)

Revision exercise 2

1 Solve the following equations by factorisation.

(a) $x^2 + 5x = 0$ **(b)** $x^2 + 7x + 6 = 0$

(c) $6x^2 - 7x - 3 = 0$ **(d)** $2x^3 - 7x^2 + 6x = 0$

2 Complete the square for

(a) $x^2 + 12x$ **(b)** $5x^2 - 12x$ **(c)** $5x^2 + 15x$

3 Solve by completing the square

$5 - 2x - 3x^2 = 0$

4 Solve the equation $x^2 + x - 9 = 0$ leaving your answers in surd form.

5 **(a)** Solve $(2x + 3)^2 = 4$

(b) Solve $2x^2 + 7x = 11$ using the formula, leaving your answers in surd form.

6 Solve the equation $5x^2 - 3 = 5x$. *E*

7 Sketch the curves with the equations

(a) $y = 6x^2 - 7x - 3$ **(b)** $y = -x^2 + 4x + 5$

8 Solve the equation $4x^2 + 4x - 7 = 0$ giving your answers in the form $p \pm q\sqrt{2}$, where p and q are real numbers to be found. *E*

9 **(a)** Solve the equation $4x^2 + 12x = 0$.

(b) $f(x) = 4x^2 + 12x + c$, where c is a constant. Given that $f(x)$ has equal roots, find the value of c and hence solve $f(x) = 0$.

10 **(a)** By completing the square find, in terms of k, the roots of the equation $x^2 + 2kx - 7 = 0$. *E*

You may need to do chapter 3 before you try this question

(b) Show that, for all values of k, the roots of $x^2 + 2kx - 7 = 0$ are real and different.

11 $f(x) = x^2 - 4x + 9$
Express $f(x)$ in the form $(x - p)^2 + q$, where p and q are constants to be found. *E*

12 Given that $f(x) = 15 - 7x - 2x^2$

(a) find the coordinates of all the points at which the graph of $y = f(x)$ crosses the coordinate axes.

(b) Sketch the graph of $y = f(x)$. *E*

13 Find the value of k for which the equation $x^2 + 10kx + 2k = 0$, where k is a constant, has real roots.

You may need to do chapter 3 before you try this question

14 Given that for all values of x

$$3x^2 + 12x + 5 \equiv p(x + q)^2 + r$$

(a) find the values of p, q and r.

(b) Hence or otherwise find minimum values of $3x^2 + 12x + 5$.

(c) Solve the equation $3x^2 + 12x + 5 = 0$. *E*

Test yourself	What to review
	If your answer is incorrect
1 Solve by factorisation **(a)** $x^2 - 3x - 10 = 0$ **(b)** $2x^3 + 3x^2 + x = 0$	*Review Heinemann Book C1 pages 15–16* *Revise for C1 page 8 Example 1*
2 Solve by completing the square $x^2 + 10x + 6 = 0$	*Review Heinemann Book C1 pages 18–19* *Revise for C1 page 8 Example 2*
3 Solve by using the formula $3x^2 - 5x + 1 = 0$	*Review Heinemann Book C1 page 20* *Revise for C1 page 9 Example 3*
4 Sketch the graph of $y = (3x + 5)(x - 4)$	*Review Heinemann Book C1 pages 21–22* *Revise for C1 page 10 Worked exam style question 2*
5 Find the values of k for which the equation $2x^2 + 5kx + k = 0$, where k is a constant, has real roots.	*Review Heinemann Book C1 page 22, Example 15 and Revise for C1 pages 9–10 Worked exam style questions 3 and 1*

Test yourself answers

1 (a) $x = -2, 5$ (b) $x = 0, -\frac{1}{2}, -1$ **2** $x = -5 \pm \sqrt{19}$ **3** $x = \frac{5 \pm \sqrt{13}}{6}$

4

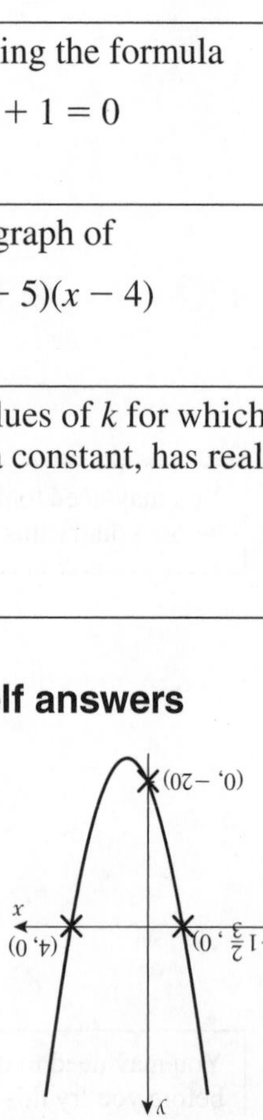

5 $k \geq \frac{8}{25}$ or $k \leq 0$

Equations and inequalities 3

Key points to remember

1 You can solve simultaneous equations by elimination or substitution.

2 You can use the substitution method to solve simultaneous equations where one equation is linear and the other is quadratic. You usually start by finding an expression for x or y from the linear equation.

3 When you multiply or divide an inequality by a negative number, you need to change the inequality sign to its opposite.

4 To solve a quadratic inequality you:

(i) solve the corresponding equation, then

(ii) sketch the graph of the quadratic function, then

(iii) use your sketch to find the required set of values.

Example 1

Solve the simultaneous equations

$$5x - 2y = 17$$
$$x - 4y = 16$$

(i) by elimination (ii) by substitution

Using **1**

(i) $10x - 4y = 34$

$x - 4y = 16$

$9x = 18$

$x = 2$

$10 - 2y = 17$

$-2y = 17 - 10$

$-2y = 7$

$y = -3\frac{1}{2}$

So the solution is $x = 2, y = -3\frac{1}{2}$

Using the method of elimination, multiply the first equation by 2 to get $4y$ in each equation.

Subtract, since the $4y$ terms have the same sign (both negative).

Use $x = 2$ in the first equation to find the value of y.

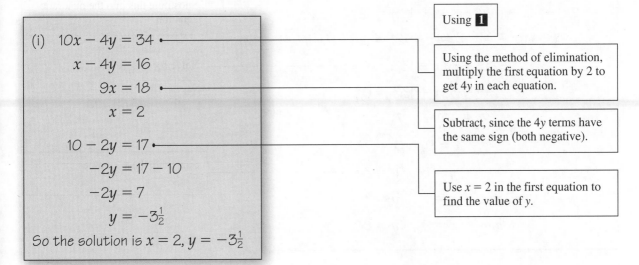

(ii) $x = 16 + 4y$

$5(16 + 4y) - 2y = 17$

$80 + 20y - 2y = 17$

$20y - 2y = 17 - 80$

$18y = -63$

$y = -3\frac{1}{2}$

$x = 16 + (4 \times -3\frac{1}{2})$

$= 16 - 14$

$= 2$

So $x = 2, y = -3\frac{1}{2}$

Using **2**

Using the substitution method, rearrange the second equation to get an expression for x.

Substitute this into the other equation.

Substitute $y = -3\frac{1}{2}$ into $x = 16 + 4y$

Worked exam style question 1

Solve the simultaneous equations

$2x - y = 11$

$xy - y^2 = 12$

$-y = 11 - 2x$

$y = 2x - 11$

$x(2x - 11) - (2x - 11)^2 = 12$

$2x^2 - 11x - (4x^2 - 44x + 121) = 12$

$2x^2 - 11x - 4x^2 + 44x - 121 = 12$

$-2x^2 + 33x - 133 = 0$

$2x^2 - 33x + 133 = 0$

$(2x - 19)(x - 7) = 0$

$x = 9\frac{1}{2}$ or $x = 7$

$y = 8$ or $y = 3$

So the solutions are $x = 7, \quad y = 3$

and $x = 9\frac{1}{2}, \quad y = 8$

Using **2**

Rearrange the linear equation to find an expression for y.

Substitute this into the quadratic equation.

Solve for x

Substitute these x-values into $y = 2x - 11$

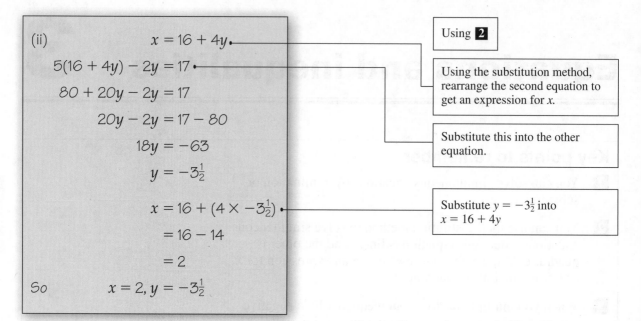

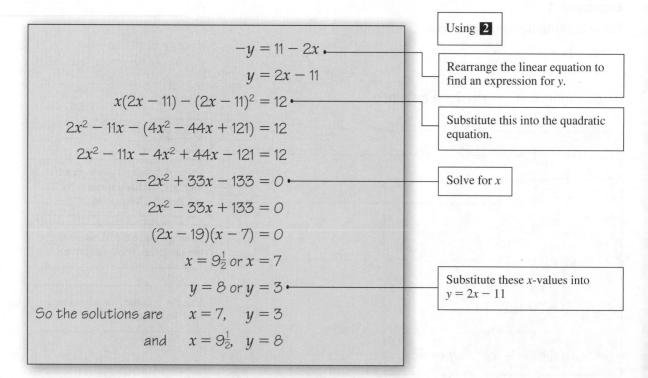

Worked exam style question 2

Find the set of values of x for which

$$5 - 3x > x + 14$$

$$-3x > x + 14 - 5$$

Subtract 5 from both sides.

$$-3x - x > 14 - 5$$

Subtract x from both sides.

$$-4x > 9$$

Dividing by -4 gives

$$x < -2\tfrac{1}{4}$$

Using **3**

Worked exam style question 3

Find the set of values of x for which

$$4x^2 + 19x - 5 > 0$$

Using **4**

$$4x^2 + 19x - 5 = 0$$

$$(4x - 1)(x + 5) = 0$$

$$x = \tfrac{1}{4} \text{ or } x = -5$$

Solve the quadratic equation.

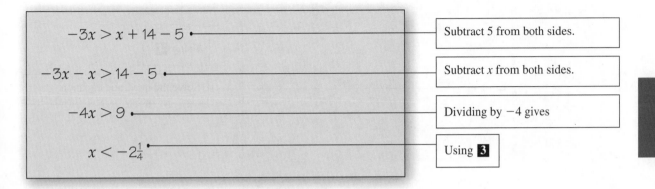

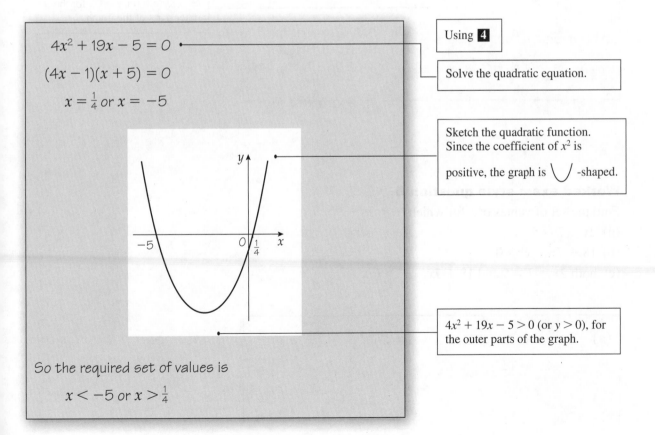

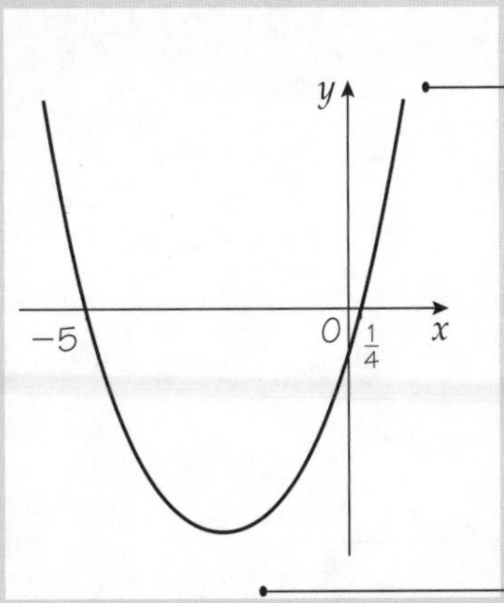

Sketch the quadratic function. Since the coefficient of x^2 is positive, the graph is $\cup$-shaped.

$4x^2 + 19x - 5 > 0$ (or $y > 0$), for the outer parts of the graph.

So the required set of values is

$$x < -5 \text{ or } x > \tfrac{1}{4}$$

Worked exam style question 4

Find the set of values of k for which $kx^2 + kx + 2 = 0$ has real roots.

For real roots

$$b^2 - 4ac \geqslant 0$$

$$k^2 - 8k \geqslant 0$$

Using **4**

$$k^2 - 8k = 0$$

Solve the quadratic equation.

$$k(k - 8) = 0$$

$$k = 0 \text{ or } k = 8$$

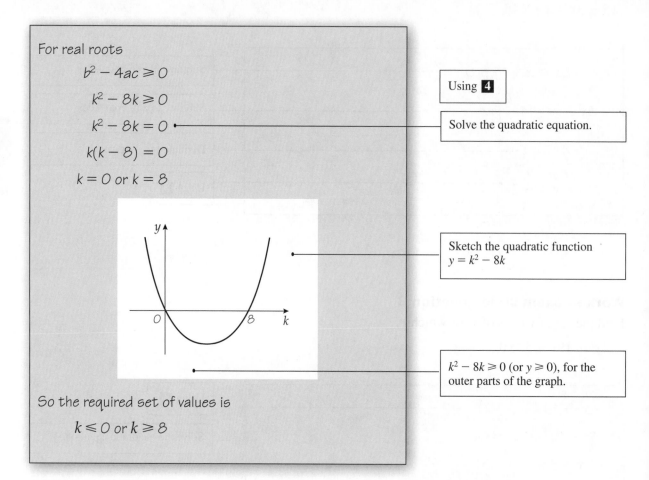

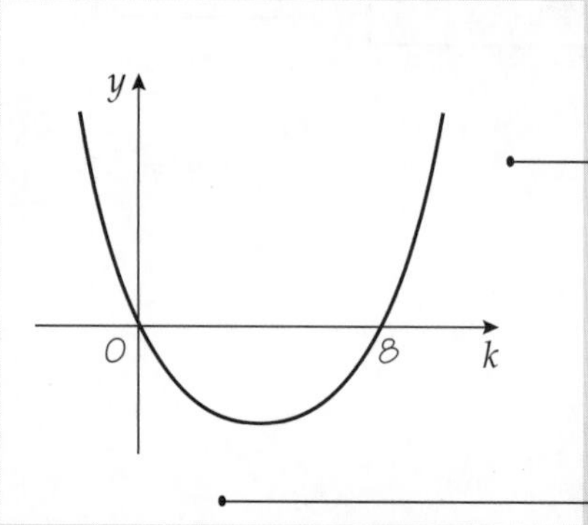

Sketch the quadratic function $y = k^2 - 8k$

$k^2 - 8k \geqslant 0$ (or $y \geqslant 0$), for the outer parts of the graph.

So the required set of values is

$$k \leqslant 0 \text{ or } k \geqslant 8$$

Worked exam style question 5

Find the set of values of x for which

(a) $2x - 7 \geqslant 5$

(b) $18 + 7x - x^2 > 0$

(c) both $2x - 7 > 5$ and $18 + 7x - x^2 > 0$.

(a) $2x - 7 \geqslant 5$

$$2x \geqslant 12$$

$$x \geqslant 6$$

(b) $18 + 7x - x^2 = 0$

Using **4**

$x^2 - 7x - 18 = 0$

Multiply the quadratic equation by -1 so it's easier to factorise.

$(x + 2)(x - 9) = 0$

$x = -2$ or $x = 9$

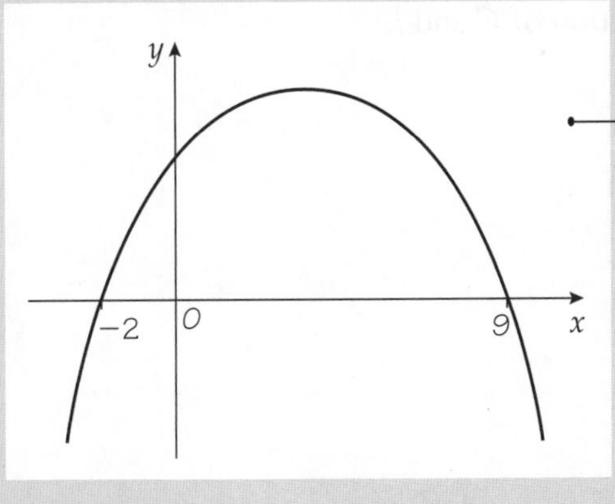

Sketch the quadratic function. $y = 18 + 7x - x^2$

So the required set of values is

$-2 < x < 9$

$18 + 7x - x^2 > 0$ for the part of the graph above the x-axis.

(c) Using the answers to **(a)** and **(b)**

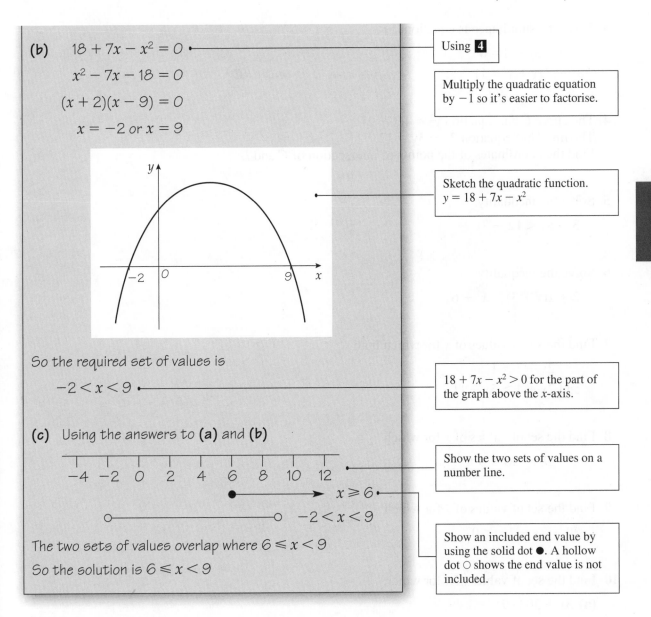

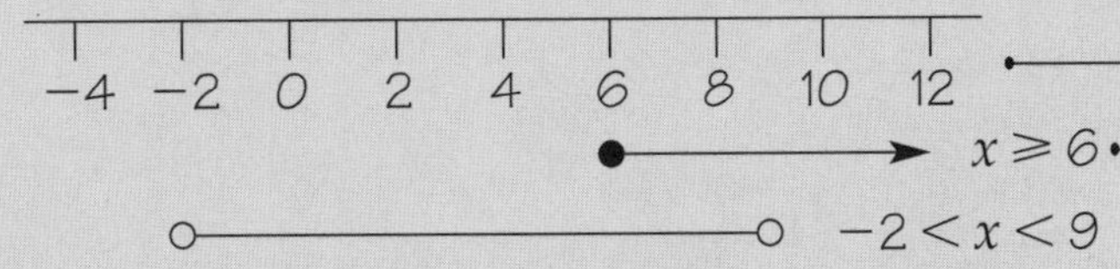

Show the two sets of values on a number line.

Show an included end value by using the solid dot ●. A hollow dot ○ shows the end value is not included.

The two sets of values overlap where $6 \leqslant x < 9$

So the solution is $6 \leqslant x < 9$

Revision exercise 3

1 Solve the simultaneous equations

$$3x - 4y = 6$$
$$x + 2y = 7$$

2 The straight line l_1 has equation $x + 3y - 4 = 0$.
The straight line l_2 has equation $3x - 5y - 19 = 0$.
Find the coordinates of the point of intersection of l_1 and l_2.

3 Solve the simultaneous equations

$$x - 3y + 1 = 0$$
$$x^2 - 3xy + y^2 = 11$$

E

4 The curve C has equation $y^2 = 4(x - 2)$.
The line l has equation $2x - 3y = 12$.
Find the coordinates of the points of intersection of C and l.

5 Solve the inequality

$$3 - 8x < 12 - 3x$$

6 Solve the inequality

$$2 + x(x + 3) > x^2 + 6$$

7 Find the set of values of x for which both

$$2x + 7 > 1$$
and $2x + 4 > 5x - 11$

8 Find the set of values of x for which

$$x^2 - 9x - 36 < 0$$

9 Find the set of values of x for which

$$30 - 7x - 2x^2 < 0$$

10 Find the set of values of x for which

(a) $3x + 30 > 9$
(b) $x^2 + x - 6 > 0$
(c) both $3x + 30 > 9$ and $x^2 + x - 6 > 0$

11 The width of a rectangular sports pitch is x m, where $x > 0$.
The length of the pitch is 20 m more than its width.
Given that the perimeter of the pitch must be less than 300 m,

(a) form a linear inequality in x.

Given that the area of the pitch must be greater than 4800 m^2,

(b) form a quadratic inequality in x.

(c) By solving your inequalities, find the set of possible values of x.

E

Test yourself

Test yourself	What to review
	If your answer is incorrect
1 Find the coordinates of the point of intersection of the lines with equations $y = 6x - 5$ and $4x - 3y = 8$	*Review Heinemann Book C1 pages 25–26* *Revise for C1 page 13 Example 1*
2 Solve the simultaneous equations $y - 2x = 5$ $y^2 - xy - x^2 = 11$	*Review Heinemann Book C1 pages 27–28* *Revise for C1 page 14 Worked exam style question 1*
3 Find the set of values of x for which $2x - x(x + 5) < x(1 - x) - 6$	*Review Heinemann Book C1 pages 29–31* *Revise for C1 page 15 Worked exam style question 2*
4 Find the set of values of x for which $2x^2 - x - 28 > 0$	*Review Heinemann Book C1 pages 32–35* *Revise for C1 page 15 Worked exam style question 3*
5 Find the set of values of x for which **(a)** $5x - 8 < 3(x - 1)$ **(b)** $3 + 2x > x^2$ **(c)** both $5x - 8 < 3(x - 1)$ and $3 + 2x > x^2$	*(a) Review Heinemann Book C1 page 29 Revise for C1 page 15 Worked exam style question 2* *(b) Review Heinemann Book C1 page 34 Revise for C1 page 15 Worked exam style question 1* *(c) Review Heinemann Book C1 page 35 Revise for C1 page 16 Worked exam style question 5*
6 Find the set of values of k for which $2x^2 + kx + 8 = 0$ has no real roots.	*Review Heinemann Book C1 pages 21–22 and 32–35* *Revise for C1 page 16 Worked exam style question 4*

Test yourself answers

1 (a) $(\frac{1}{2}, -2)$ **2** $x = -1, y = 3$ or $x = -14, y = -23$ **3** $x > 1\frac{1}{2}$ **4** $x < -3\frac{1}{2}$ or $x > 4$

5 (a) $x < 2\frac{1}{2}$ **(b)** $-1 < x < 3$ **(c)** $-1 < x < 2\frac{1}{2}$ **6** $-8 < k < 8$

Sketching curves 4

Key points to remember

1 You should know the shapes of the following curves:

(i) $y = x^2$

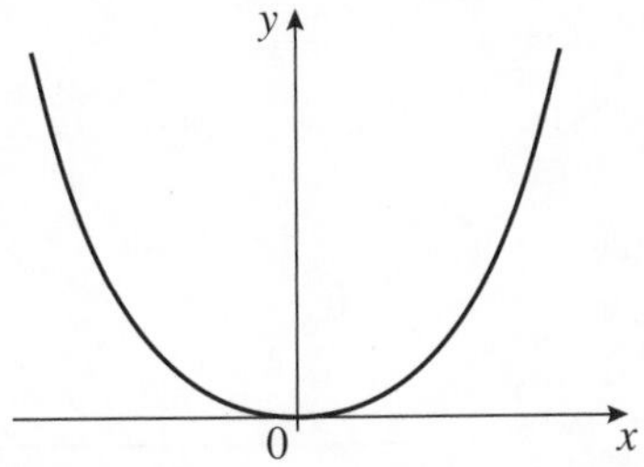

(ii) $y = x^3$

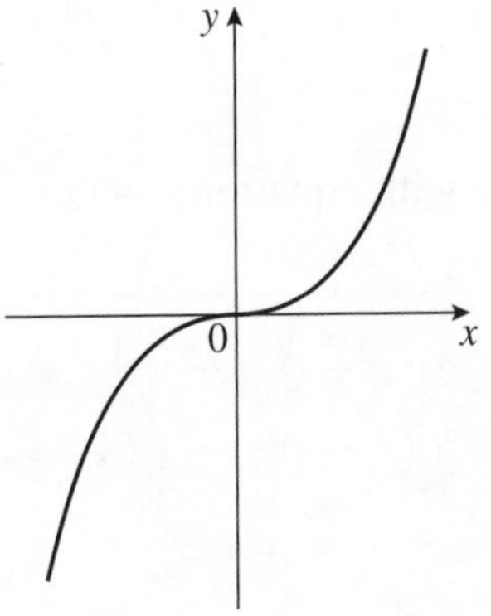

(iii) $y = (x - a)(x - b)(x - c)$

Put $y = 0 \Rightarrow x = a, b, c$

$x = 0 \Rightarrow y = -abc$

$x \to \infty, y \to \infty$

$x \to -\infty, y \to -\infty$

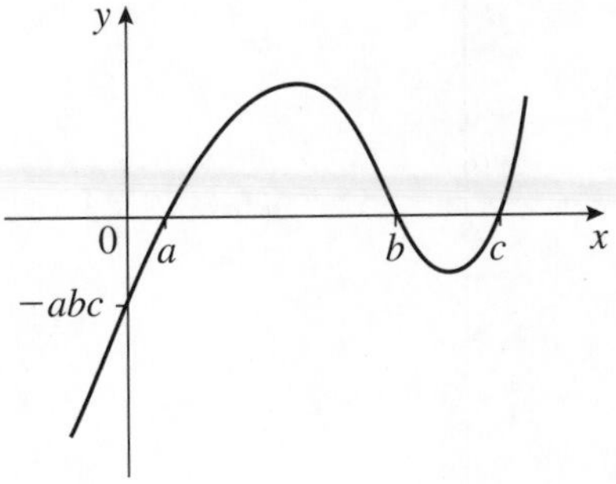

(iv) $y = \dfrac{k}{x}, k > 0$

$x = 0$ and $y = 0$ are asymptotes

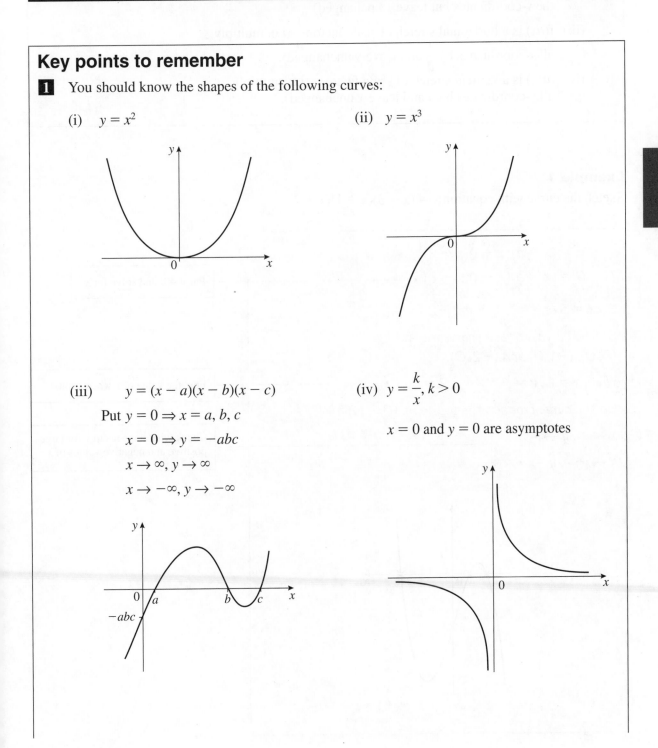

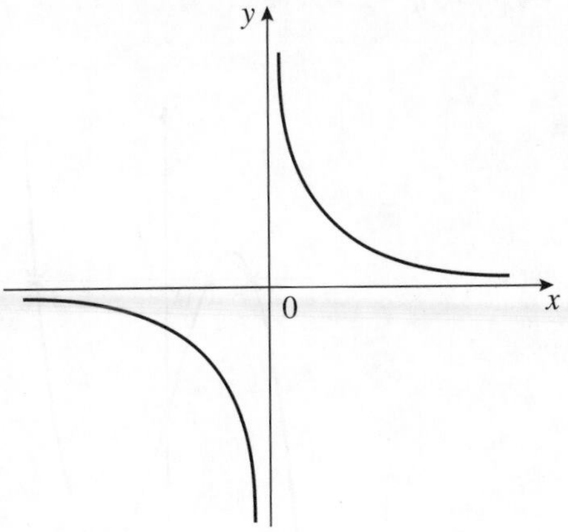

2 You should know how to use the following transformations:

(i) $f(x + a)$ is a horizontal translation of $-a$ (i.e. subtract a from all x-coordinates but leave y-coordinates unchanged).

(ii) $f(x) + a$ is a vertical translation of $+a$ (i.e. add a to all the y-coordinates but leave x unchanged).

(iii) $f(ax)$ is a horizontal stretch of scale factor $\frac{1}{a}$ (i.e. multiply all x-coordinates by $\frac{1}{a}$ and leave y unchanged).

(iv) $af(x)$ is a vertical stretch of scale factor a (i.e. multiply all y-coordinates by a and leave x unchanged).

Example 1

Sketch the curve with equation $y = (x - 3)(x + 1)(x + 2)$

$y = (x - 3)(x + 1)(x + 2)$

$0 = (x - 3)(x + 1)(x + 2)$ — Put $y = 0$ and solve for x

So $x = 3$ or $x = -1$ or $x = -2$

So the curve crosses the x-axis at $(3, 0)$, $(-1, 0)$ and $(-2, 0)$.

When $x = 0$, $y = -3 \times 1 \times 2 = -6$ — Find the value of y when $x = 0$

So the curve crosses the y-axis at $(0, -6)$.

$x \to \infty, y \to \infty$

$x \to -\infty, y \to -\infty$

— Check what happens to y for large positive and negative values of x

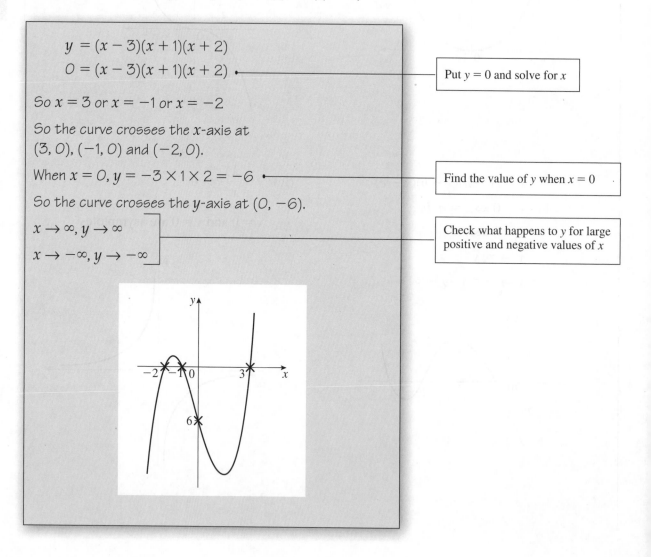

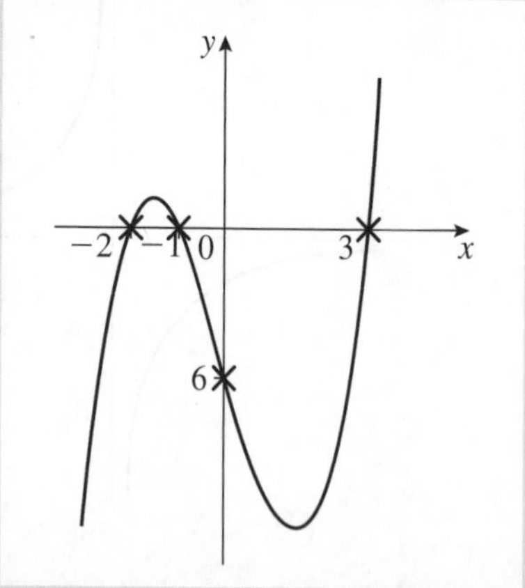

Example 2

Sketch the following curves and indicate the points where the curves cross the coordinate axes.

(a) $y = 2(x + 3)(x - 2)^2$ **(b)** $y = 2x^3 - 5x^2 - 3x$

(a) $y = 2(x + 3)(x - 2)^2$

$0 = 2(x + 3)(x - 2)^2$ — Put $y = 0$ and solve for x

So $x = -3$ or 2 (twice)

So the curve crosses the x-axis at $(-3, 0)$ and $(2, 0)$

When $x = 0$, $y = 2 \times 3 \times (-2)^2 = 24$ — Find the value of y when $x = 0$

So the curve crosses the y-axis at $(0, 24)$.

$x \to \infty, y \to \infty$

$x \to -\infty, y \to -\infty$

— Check what happens to y for large positive and negative values of x

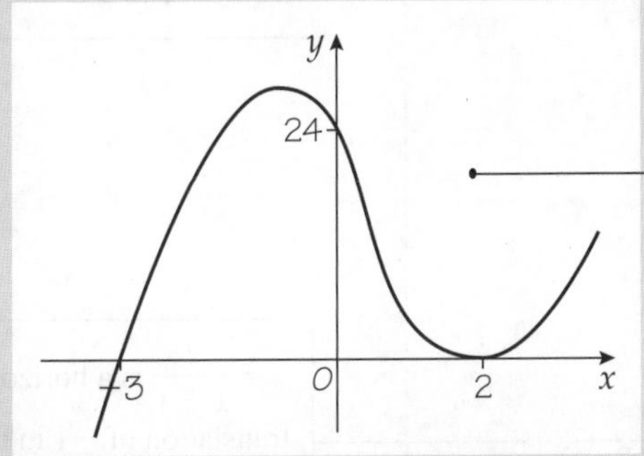

Note the double root at $x = 2$ means the curve touches the x-axis here

(b) $2x^3 - 5x^2 - 3x = x(2x^2 - 5x - 3)$ — First factorise

$= x(2x + 1)(x - 3)$

$0 = x(2x + 1)(x - 3)$ — Put $y = 0$ and solve for x

$x = 0, -\frac{1}{2}, 3$

So the curve crosses the axes at $(0, 0)$, $(-\frac{1}{2}, 0)$ and $(3, 0)$

$x \to \infty, y \to \infty$

$x \to -\infty, y \to -\infty$

— Check what happens to y for large positive and negative values of x

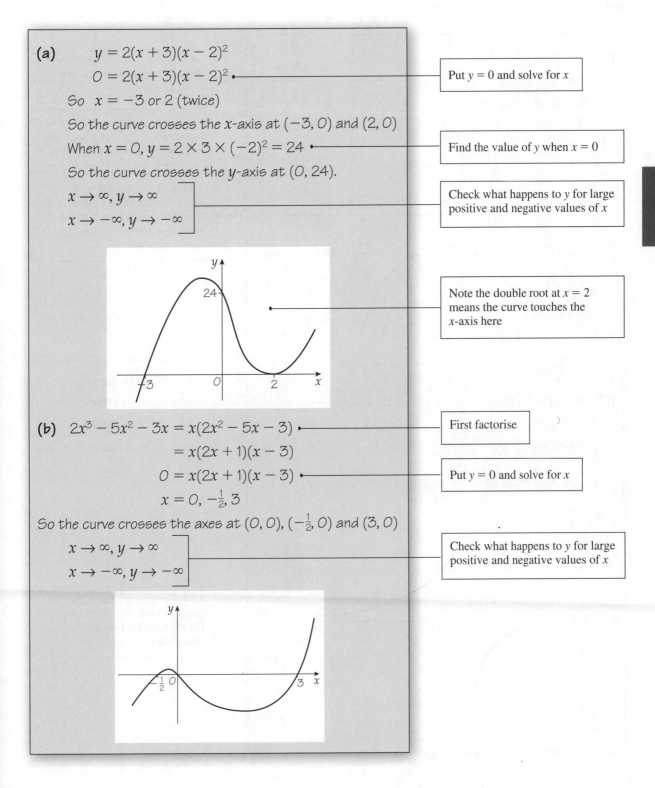

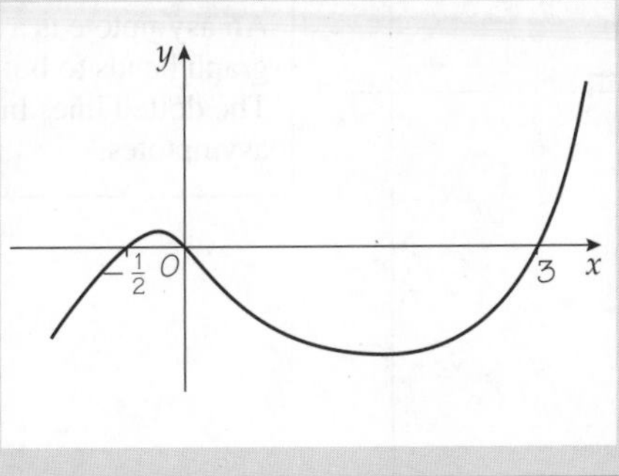

Worked exam style question 1

Sketch the following curves and indicate coordinates of points where they cross the axes. In **(b)** state the equations of any asymptotes.

(a) $y = (4 - x)^3$ **(b)** $y = \dfrac{2}{x + 1}$

(a) $y = (4 - x)^3$

$\Rightarrow y = [-(x - 4)]^3$

$\Rightarrow y = -(x - 4)^3$

If $f(x) = -x^3$ then $-(x - 4)^3$ is $f(x - 4)$.

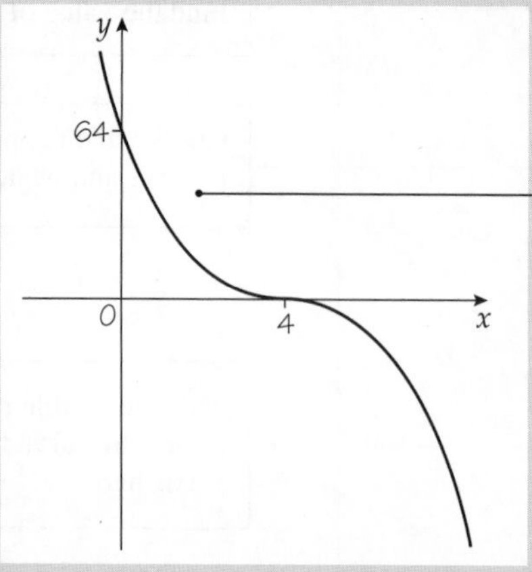

Write the equation in the form $y = \pm(x - a)^3$

So this curve will have the same shape as $y = -x^3$ but it has been translated +4 in the x-direction

Alternatively, solve $0 = (4 - x)^3$ to see that the curve crosses at (4, 0) and when $x = 0$, $y = 4^3 = 64$. Then $x \to \infty$, $y \to -\infty$ and $x \to -\infty$, $y \to +\infty$

(b) If $f(x) = \dfrac{2}{x}$ then $\dfrac{2}{x + 1} = f(x + 1)$

When $x = 0$, $y = \dfrac{2}{0 + 1} = 2$

The asymptotes are $y = 0$ and $x = -1$

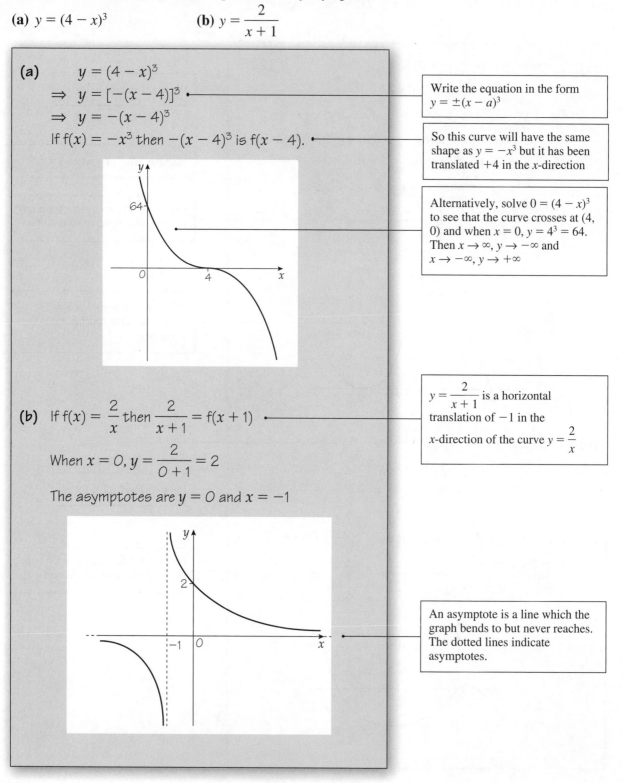

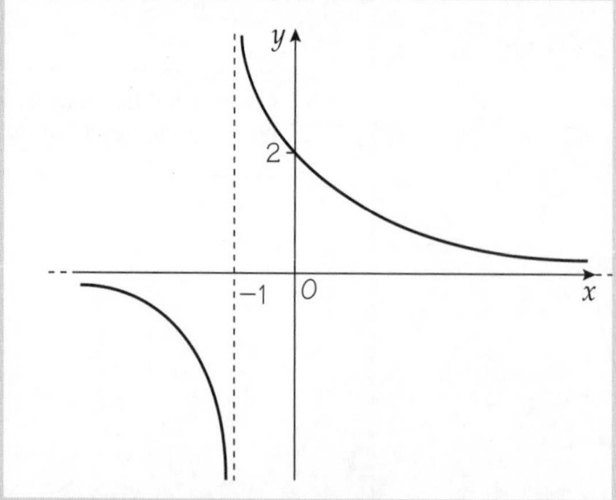

$y = \dfrac{2}{x + 1}$ is a horizontal translation of -1 in the x-direction of the curve $y = \dfrac{2}{x}$

An asymptote is a line which the graph bends to but never reaches. The dotted lines indicate asymptotes.

Worked exam style question 2

(a) On the same axes, sketch the curves with equations

$y = \dfrac{2}{1-x}$ and $y = 2 - x^2$

Make and label the points of intersection.

(b) Use algebra to find the coordinates of the points of intersection.

(c) State the solutions to the equation $(x^2 - 2)(x - 1) = 2$.

(a) If $f(x) = \dfrac{-2}{x}$ then $\dfrac{-2}{x-1} = f(x-1)$

> $y = \dfrac{2}{1-x} = \dfrac{-2}{x-1}$ which has the same shape as $y = \dfrac{-2}{x}$, but has been translated by +1 in the x-direction.

When $x = 0$, $y = \dfrac{2}{1-0} = 2$

The asymptotes are $y = 0$ and $x = 1$.

If $f(x) = -x^2$ then $2 - x^2 = -x^2 + 2$ is $f(x) + 2$

> $y = 2 - x^2$ is the same shape as $y = -x^2$ but it has been translated by +2 in the y-direction.

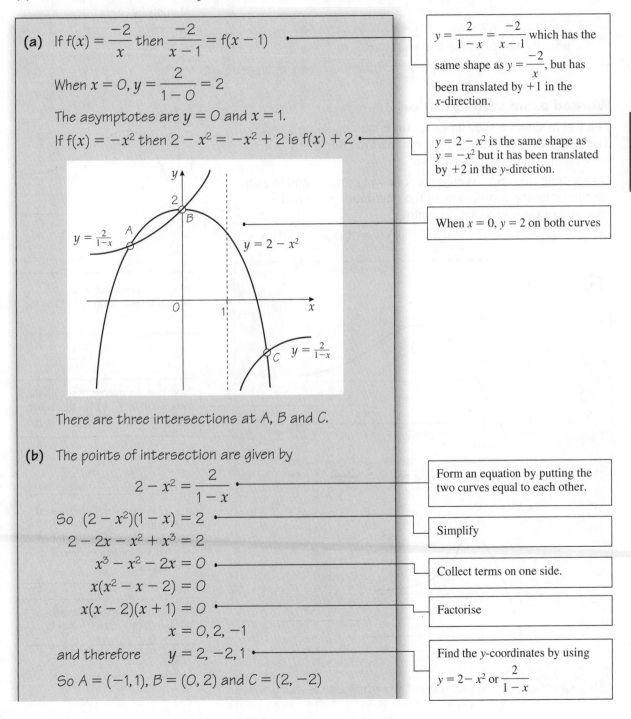

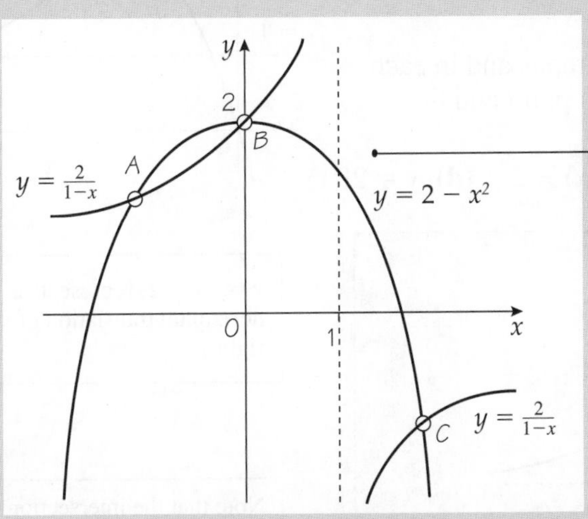

> When $x = 0$, $y = 2$ on both curves

There are three intersections at A, B and C.

(b) The points of intersection are given by

$$2 - x^2 = \frac{2}{1-x}$$

> Form an equation by putting the two curves equal to each other.

So $(2 - x^2)(1 - x) = 2$

> Simplify

$$2 - 2x - x^2 + x^3 = 2$$

$$x^3 - x^2 - 2x = 0$$

> Collect terms on one side.

$$x(x^2 - x - 2) = 0$$

$$x(x - 2)(x + 1) = 0$$

> Factorise

$$x = 0, 2, -1$$

and therefore $y = 2, -2, 1$

> Find the y-coordinates by using $y = 2 - x^2$ or $\dfrac{2}{1-x}$

So $A = (-1, 1)$, $B = (0, 2)$ and $C = (2, -2)$

(c) The graphs intersect when

$$\frac{2}{1-x} = 2 - x^2$$

or $$\frac{2}{x-1} = x^2 - 2$$

Multiply by -1

so $(x^2 - 2)(x - 1) = 2$

Multiply by $(x - 1)$

The graphs intersect at $x = 0$, 2 and -1 so these are the solutions to the equation.

Worked exam style question 3

The figure shows a sketch of the curve with equation $y = \text{f}(x)$. The point (6, 3) is a maximum point and the line $y = 1$ is a horizontal asymptote to the curve.

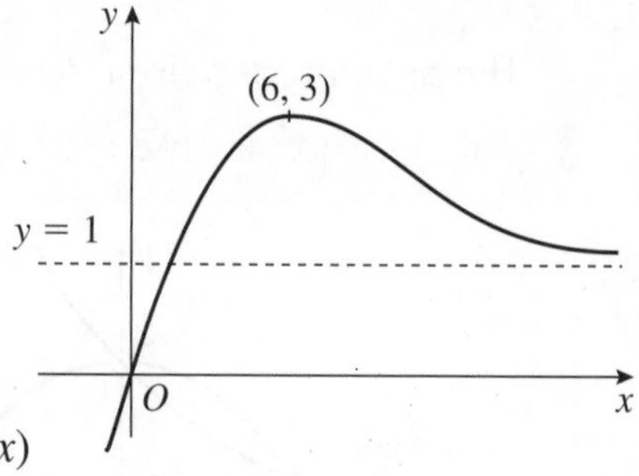

On separate diagrams, sketch the following graphs and in each case indicate the coordinates of the maximum point and the equation of the horizontal asymptote.

(a) $y = \text{f}(x - 2)$ **(b)** $y = \text{f}(2x)$ **(c)** $y = \text{f}(x) - 2$ **(d)** $y = 2\text{f}(x)$

(a)

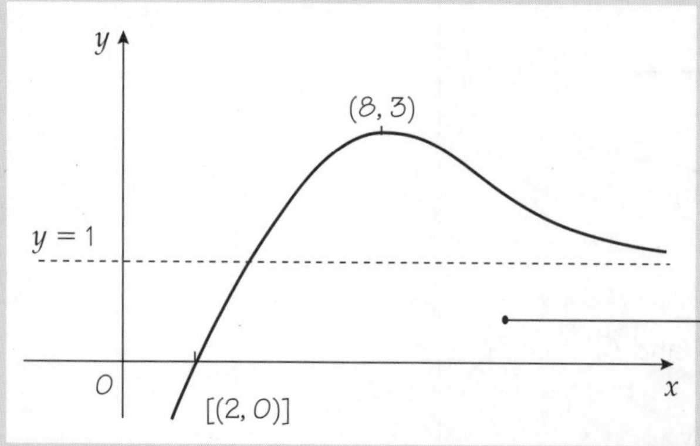

$y = \text{f}(x - 2)$ represents a horizontal translation of +2 to the right

Note that the intersection with the x-axis will be (2, 0) but this is not required here

The maximum point is (8, 3) and the asymptote is still $y = 1$.

(b)

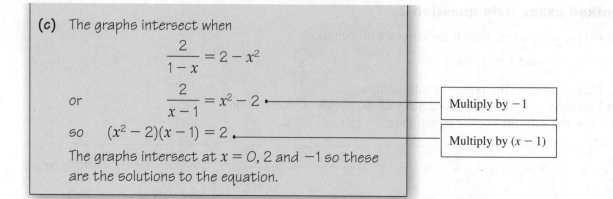

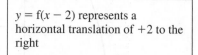

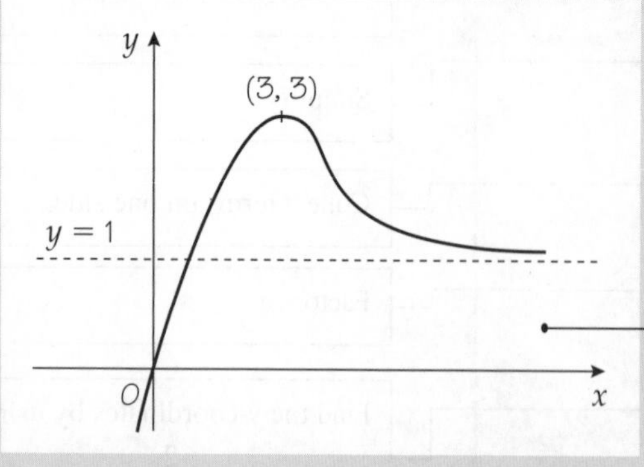

$y = \text{f}(2x)$ represents a horizontal translation of $\times \frac{1}{2}$.
All x-coordinates are multiplied by $\frac{1}{2}$ but y-coordinates remain unchanged

Note that the curve still passes through (0, 0) since $0 \times \frac{1}{2} = 0$

The maximum point is (3, 3) and the asymptote is still $y = 1$.

(c)

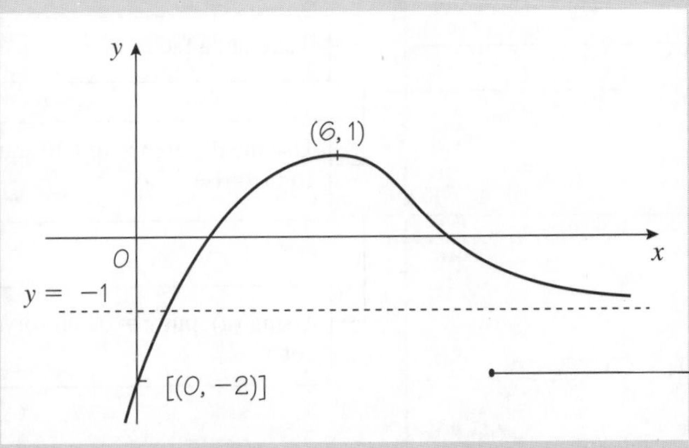

$y = f(x) - 2$ represents a vertical translation of -2. This means that the graph moves 2 units down.

Note that the curve will cross the y-axis at $(0, -2)$

The horizontal asymptote is $y = -1$ and the maximum point is $(6, 1)$.

(d)

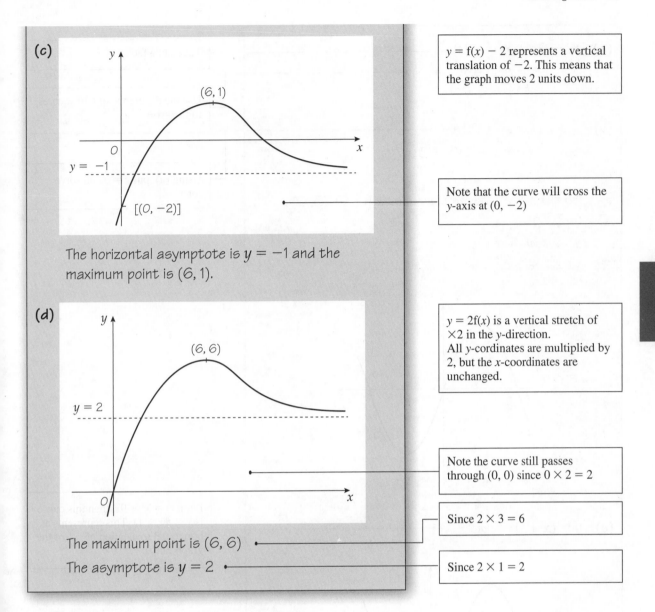

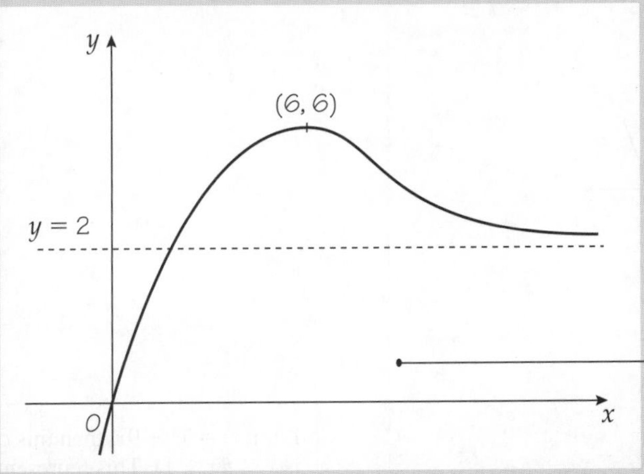

$y = 2f(x)$ is a vertical stretch of $\times 2$ in the y-direction. All y-cordinates are multiplied by 2, but the x-coordinates are unchanged.

Note the curve still passes through $(0, 0)$ since $0 \times 2 = 2$

The maximum point is $(6, 6)$

Since $2 \times 3 = 6$

The asymptote is $y = 2$

Since $2 \times 1 = 2$

Worked exam style question 4

(a) Factorise completely $x^3 - 9x$

(b) Sketch the curve with equation $y = x^3 - 9x$, showing the coordinates of the points where the curve crosses the x-axis.

(c) On a separate diagram, sketch the curve with equation

$$y = (x + 1)^3 - 9(x + 1)$$

showing the coordinates of the points where the curve crosses the x-axis.

(a) $x^3 - 9x = x(x^2 - 9)$

$= x(x - 3)(x + 3)$

Take out a factor of x

Use the difference of two squares to factorise.

(b) $0 = x^3 - 9x$

So $0 = x(x - 3)(x + 3)$

$x = 0, 3$ or -3

So the curve crosses the axes at $(0, 0)$, $(3, 0)$ and $(-3, 0)$

$x \to \infty, y \to \infty$

$x \to -\infty, y \to -\infty$

Using **(a)**, put $y = 0$ and solve for x

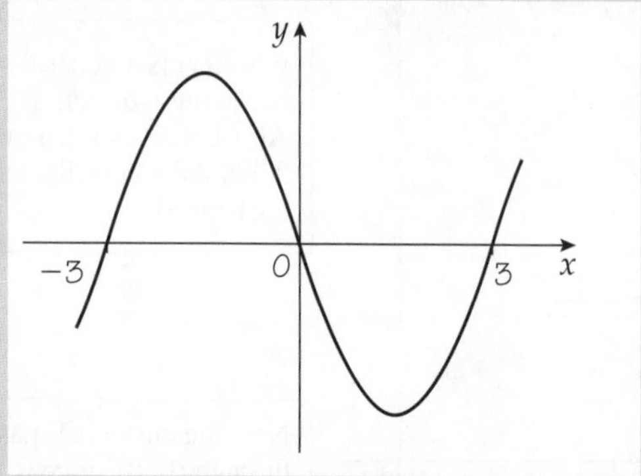

(c) $y = (x + 1)^3 - 9(x + 1)$

Let $f(x) = x^3 - 9x$, then this curve is $y = f(x + 1)$. This represents a horizontal translation of -1 of the graph in **(b)**.

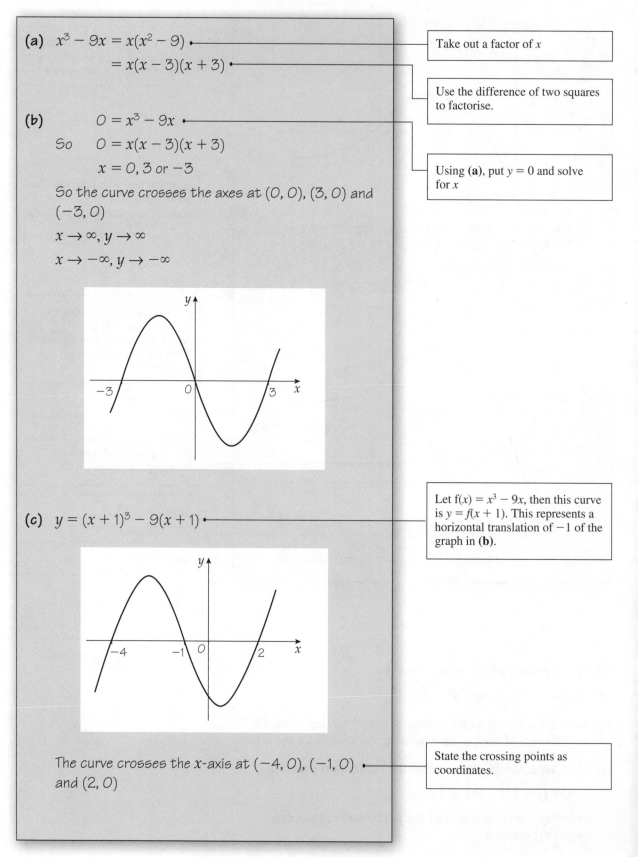

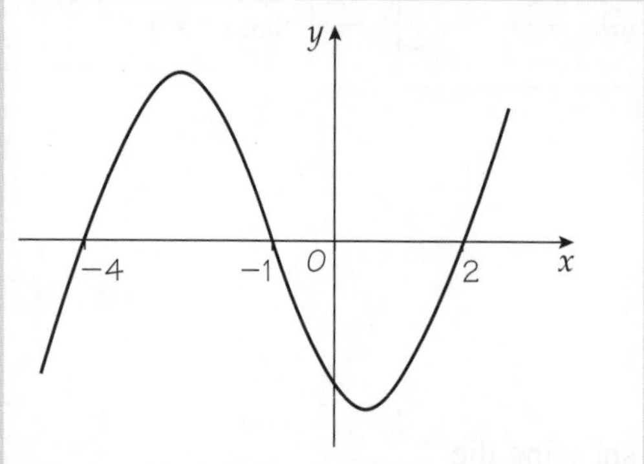

The curve crosses the x-axis at $(-4, 0)$, $(-1, 0)$ and $(2, 0)$

State the crossing points as coordinates.

Revision exercise 4

1 The curve C has equation $y = 2x^3 + x^2 - x$

(a) Factorise $2x^3 + x^2 - x$ **(b)** Sketch C

2 **(a)** Factorise $5x^2 - x^3 - 6x$

(b) Sketch the curve with equation $y = 5x^2 - x^3 - 6x$

3 The curve C has equation $y = x(x^2 - 4)$ and the straight line L has equation $y - 5x = 0$

(a) Sketch C and L on the same axes.

(b) Write down the coordinates of the points at which C meets the coordinate axes.

(c) Using algebra, find the coordinates of the points at which L intersects C.

4 **(a)** Sketch the curve with equation $y = 4x^2 - 3x^3$

(b) On the same axes, sketch the line with equation $y = x$

(c) Use algebra to find the coordinates of the points where $y = x$ crosses the curve.

5 The curve C_1 has equation $y = (x - 1)(x + 3)$ and the curve C_2 has equation $y = x(7 - x)$.

(a) On the same axes, sketch the graphs of C_1 and C_2. The curves C_1 and C_2 meet at the points A and B.

(b) Find the coordinates of points A and B.

6 On separate diagrams, sketch the curves with equations

(a) $y = x^2 - 1$ **(b)** $y = (x - 2)^2 - 1$ **(c)** $y = 1 - x^2$

In each part, show clearly the coordinates of any point at which the curve meets the x-axis or the y-axis.

7 On separate diagrams, sketch the curves with equations

(a) $y = -\dfrac{3}{x}$ $\quad -3 \leqslant x \leqslant 3,\ x \neq 0$

(b) $y = 1 - \dfrac{3}{x}$ $\quad -3 \leqslant x \leqslant 3,\ x \neq 0$

(c) $y = \dfrac{-3}{x+2}$ $\quad -3 \leqslant x \leqslant 3,\ x \neq -2$

8 The figure shows a sketch of the curve with equation $y = f(x)$. The curve has a maximum point at $(0, 0)$ and passes through the point $(4, 0)$.

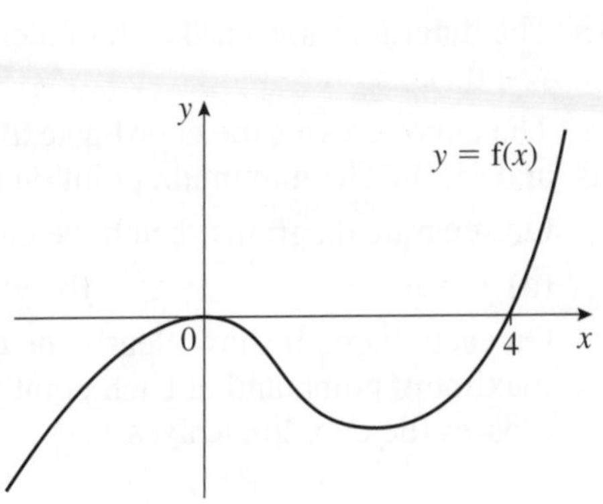

On separate diagrams, sketch

(a) $y = f(-x)$

(b) $y = -f(x)$

(c) $y = f(x + 4)$

In each case, indicate the coordinates of points at which the curve crosses the x-axis and the y-axis.

9 The figure shows a sketch of the curve with equation $y = \mathrm{f}(x)$ where $\mathrm{f}(x) = x^3 + ax^2 + bx$ and a, b and c are integers. The curve passes through the points (0, 0), (3, 0) and (5, 0).

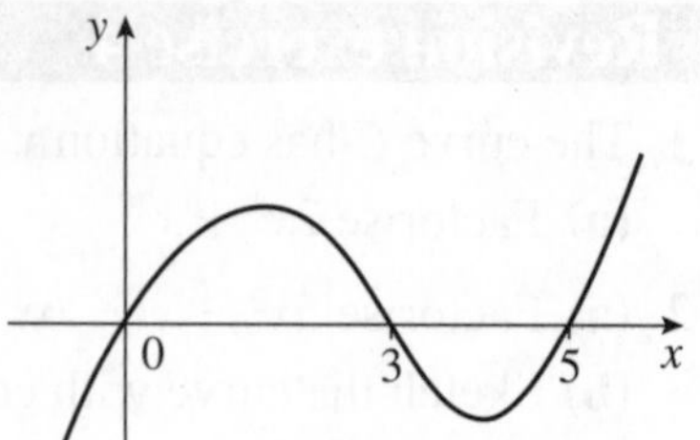

(a) Find the value of a and the value of b.

The graph of $y = k\mathrm{f}(x)$ passes through the point (1, 2).

(b) Find the value of k. **E**

10 The curve C_1, passes through the point (2, 2) and has equation $y = x^2(a - x)$ where a is a positive constant.

(a) Find the value of a. **(b)** Sketch C_1.

The curve C_2 has equation $y = x^2(a - x) + b$, where a has the value from **(a)** and b is a constant.

Given that C_2 passes through the point (1, 3)

(c) find the value of b.

(d) Write down the coordinates of the minimum point of C_2. **E**

11 On the same axes, sketch the graphs of

(a) $y = x^3 + 1$

(b) $y = 1 - x^3$

(c) $y = (1 - x)^3$

For each graph, show clearly the coordinates of any point at which the curve meets the x-axis or the y-axis. **E**

12 The figure shows a sketch of a curve with equation $y = \mathrm{f}(x)$. The curve passes through the origin O and the lines $y = 2$ and $x = 1$ are asymptotes.

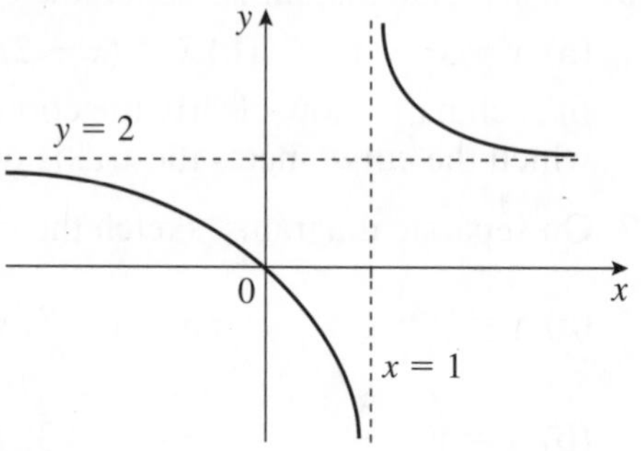

On separate diagrams sketch

(a) $y = \mathrm{f}(x + 2)$

(b) $y = -\mathrm{f}(x)$

(c) $y = \mathrm{f}(-x)$

In each case, state the coordinates of any points where the curves cross the x-axis and state the equations of any asymptotes. **E**

13 The diagram shows a sketch of a curve with equation $y = \mathrm{f}(x)$.

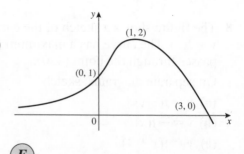

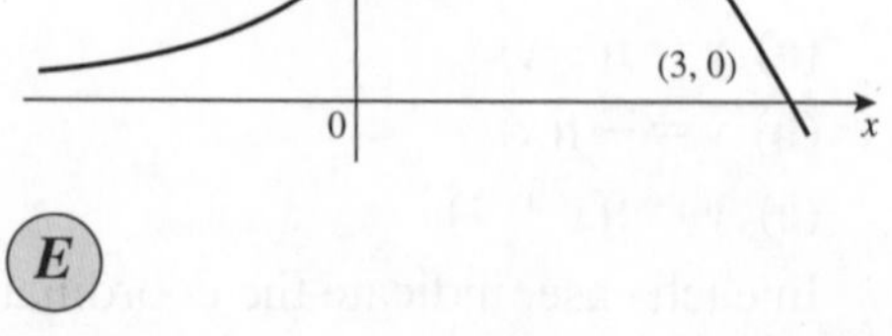

The curve crosses the coordinate axes at the points (0, 1) and (3, 0). The maximum point on the curve is (1, 2).

On separate diagrams sketch the curve with equation

(a) $y = \mathrm{f}(x + 1)$ **(b)** $y = \mathrm{f}(2x)$

On each diagram show clearly the coordinates of the maximum point, and of each point at which the curve crosses the coordinate axes. **E**

Test yourself

What to review

If your answer is incorrect

1 The curve C has equation $y = 2x^3 + x^2 - 6x$

(a) Factorise $2x^3 + x^2 - 6x$

(b) Sketch C.

Review Heinemann Book C1 pages 38–43
Revise for C1 page 24
Worked exam style question 1

---

2 **(a)** On the same axes, sketch the graphs of

(i) $y = (x - 1)^3$

(ii) $y = \frac{2}{x}$

(b) State the number of solutions to the equation $x(x - 1)^3 = 2$

Review Heinemann Book C1 pages 43–51
Revise for C1 page 25
Worked exam style question 2

---

3 The curve C has equation $y = \text{f}(x)$ where $\text{f}(x) = x(x - 4)$.
On separate axes, sketch the curves with equations

(a) $y = \text{f}(2x)$ **(b)** $y = \text{f}(-x)$

(c) $y = \text{f}(x) + 4$ **(d)** $y = \text{f}(x - 1)$

In each case, mark the coordinates of the points where the curve meets the coordinate axes.

Review Heinemann Book C1 pages 51–59
Revise for C1 page 26
Worked exam style question 3

4

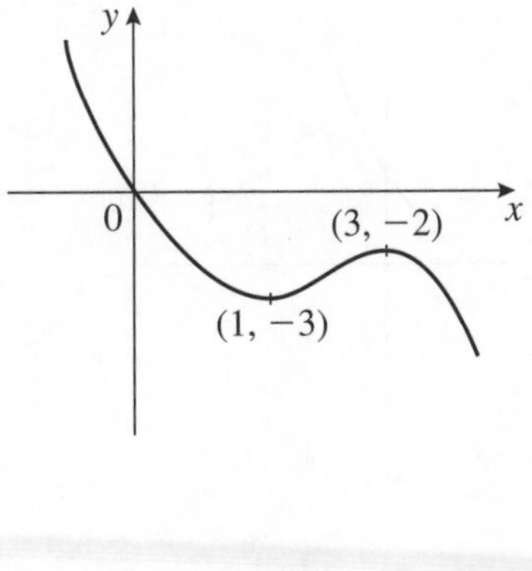

Review Heinemann Book C1 pages 60–61
Revise for C1 page 26
Worked exam style question 3

The diagram shows a sketch of the curve with equation $y = \text{f}(x)$.
The curve passes through the origin and has a minimum point at $(1, -3)$ and a maximum point at $(3, -2)$.

On separate diagrams, sketch the curves with equations

(a) $y = \text{f}(x) + 1$ **(b)** $y = \text{f}(x + 1)$ **(c)** $y = \text{f}(3x)$

On each sketch, mark the coordinates of the turning points and the y-coordinate of the point where the curve meets the y-axis.

Test yourself answers

1 **(a)** $x(2x-3)(x+2)$ **(b)**

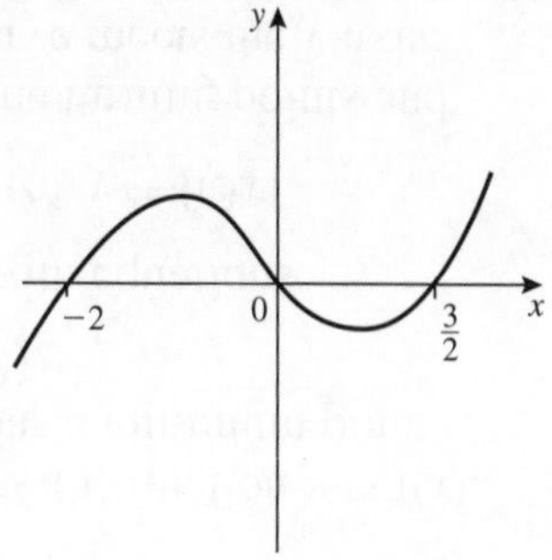

2 **(a)**

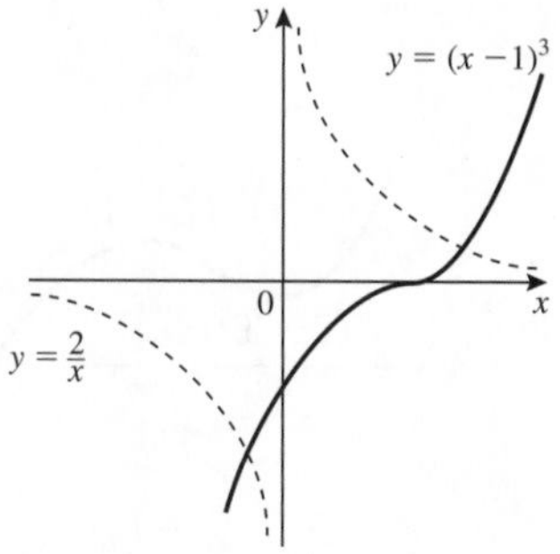

(b) 2 solutions

3 **(a)**

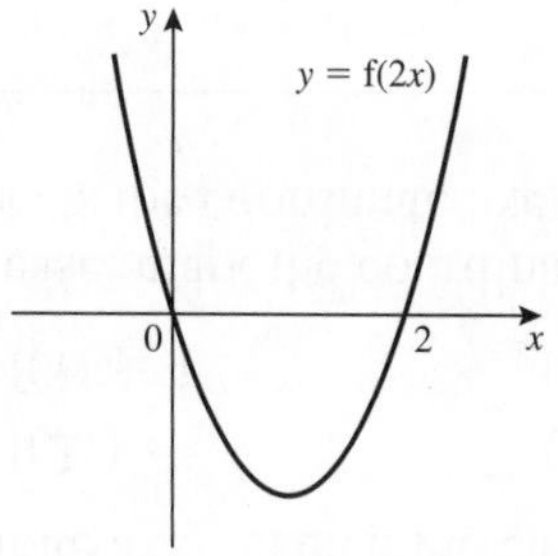

(b)

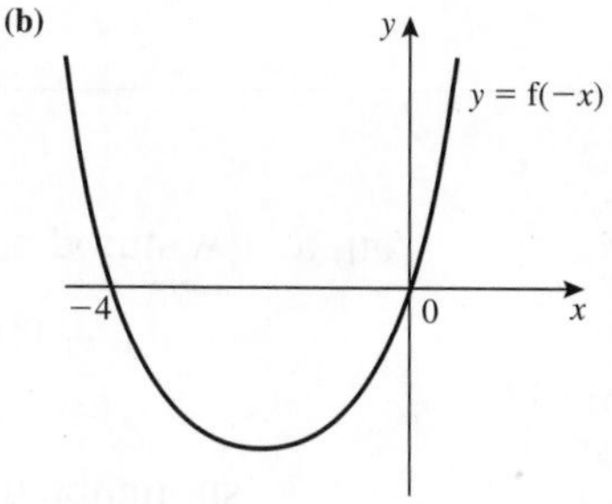

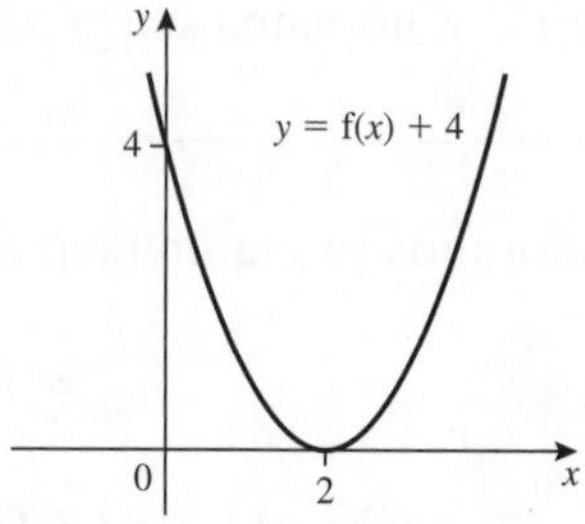

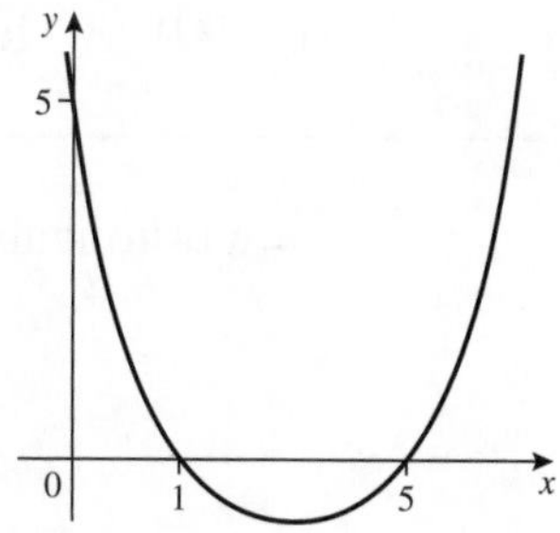

4 **(a)**

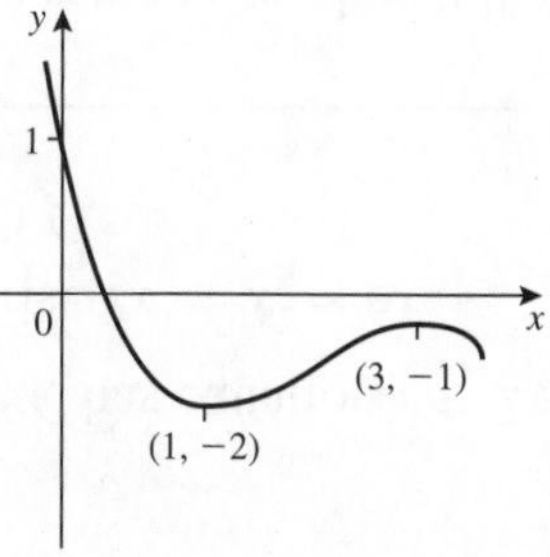

(b)

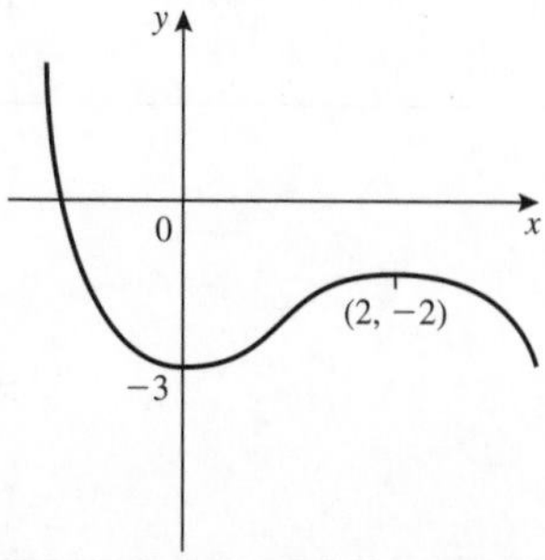

(c)

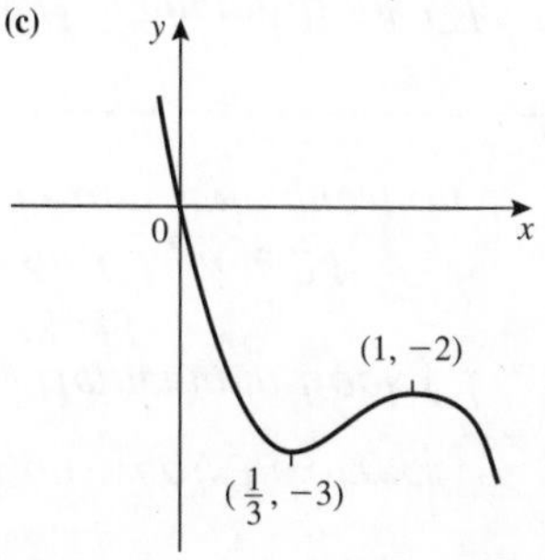

Coordinate geometry in the (*x*, *y*) plane

5

Key points to remember

1 • In the general form

$$y = mx + c,$$

where m is the gradient and $(0, c)$ is the intercept on the y-axis.

• In the general form

$$ax + by + c = 0,$$

where a, b and c are integers.

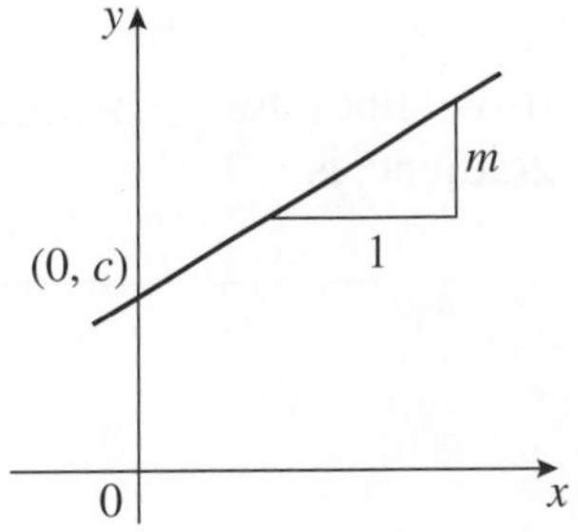

2 You can work out the gradient m of the line joining the point with coordinates (x_1, y_1) to the point with coordinates (x_2, y_2) by using the formula:

$$m = \frac{y_2 - y_1}{x_2 - x_1}$$

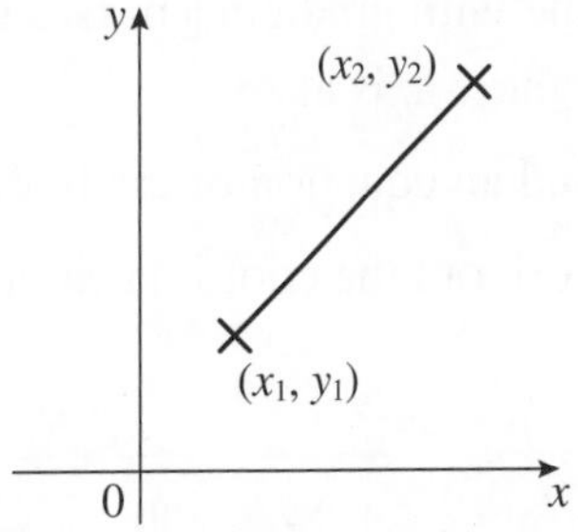

3 You can find the equation of a line with gradient m that passes through the point with coordinates (x_1, y_1) by using the formula:

$$y - y_1 = m(x - x_1)$$

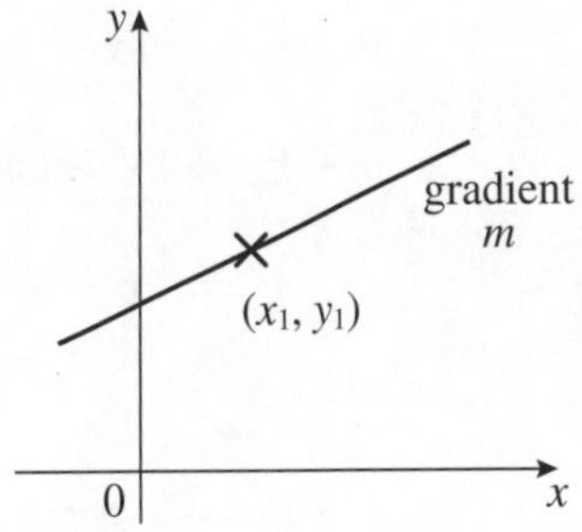

4 You can find the equation of the line that passes through the points with coordinates (x_1, y_1) and (x_2, y_2) by using the formula:

$$\frac{y - y_1}{y_2 - y_1} = \frac{x - x_1}{x_2 - x_1}$$

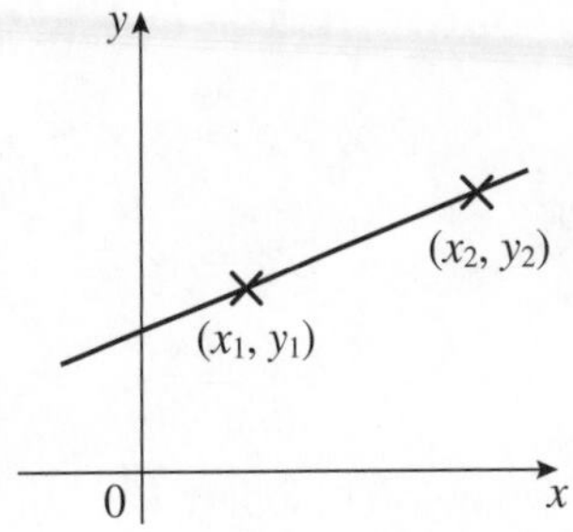

5 If a line has a gradient m, a line perpendicular to it has a gradient of $\frac{-1}{m}$.

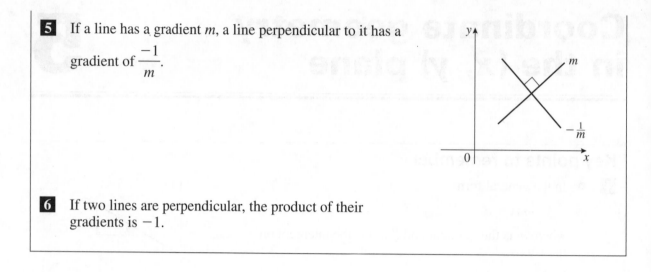

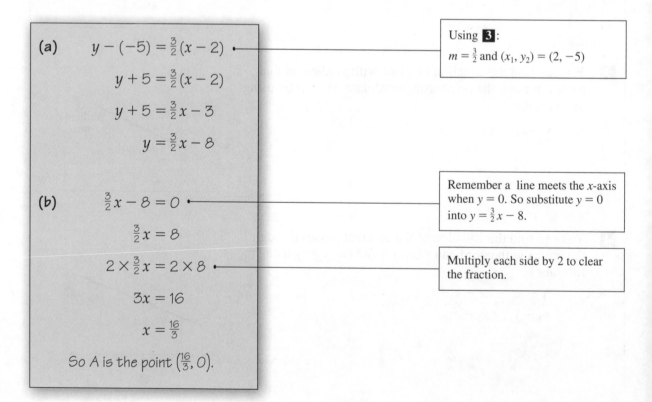

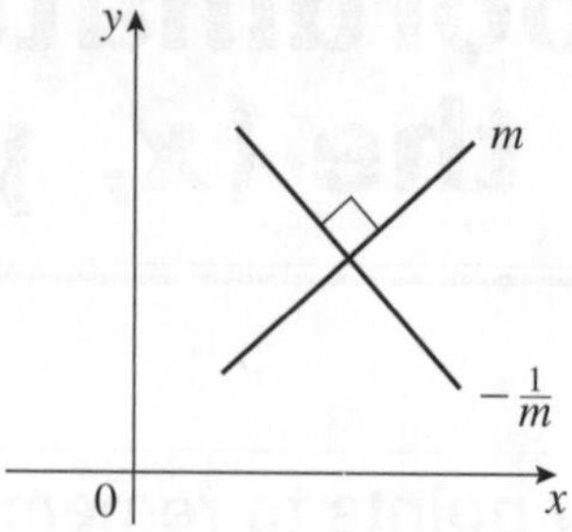

6 If two lines are perpendicular, the product of their gradients is -1.

Example 1

The line with gradient $\frac{3}{2}$ passes through the point $(2, -5)$ and meets the x-axis at A.

(a) Find an equation of the line.

(b) Work out the coordinates of A.

(a) $y - (-5) = \frac{3}{2}(x - 2)$

Using **3**: $m = \frac{3}{2}$ and $(x_1, y_2) = (2, -5)$

$y + 5 = \frac{3}{2}(x - 2)$

$y + 5 = \frac{3}{2}x - 3$

$y = \frac{3}{2}x - 8$

(b) $\frac{3}{2}x - 8 = 0$

Remember a line meets the x-axis when $y = 0$. So substitute $y = 0$ into $y = \frac{3}{2}x - 8$.

$\frac{3}{2}x = 8$

$2 \times \frac{3}{2}x = 2 \times 8$

Multiply each side by 2 to clear the fraction.

$3x = 16$

$x = \frac{16}{3}$

So A is the point $(\frac{16}{3}, 0)$.

Example 2

Show the lines $4x - 5y - 10 = 0$ and $y = -\frac{5}{4}x + 7$ are perpendicular.

$4x - 5y - 10 = 0$

$5y = 4x - 10$

Rearrange the equation into the form $y = mx + c$

$\frac{5y}{5} = \frac{4}{5}x - \frac{10}{5}$

Divide each term by 5 and simplify.

$y = \frac{4}{5}x - 2$

So the gradient of $4x - 5y - 10 = 0$ is $\frac{4}{5}$

Using **1**: comparing $y = \frac{4}{5}x - 2$ with $y = mx + c$, $m = \frac{4}{5}$

and the gradient of $y = -\frac{5}{4} + 7$ is $-\frac{5}{4}$

$\frac{4}{5} \times -\frac{5}{4} = -1$

so the lines are perpendicular.

Using **6**: the product of the gradients of perpendicular line is -1

Worked exam style question 1

Find an equation of the line that passes through the point $(-4, -5)$ and is perpendicular to the line $y = -\frac{2}{7}x + 3$. Write your answer in the form $ax + by + c = 0$, where a, b and c are integers.

The gradient of this line is $-\frac{2}{7}$,

Using **1**: comparing $y = -\frac{2}{7}x + 3$ with $y = mx + c$, $m = -\frac{2}{7}$

so the gradient of the perpendicular line is $\frac{7}{2}$.

Using **5**: here $m = -\frac{2}{7}$, so $-\frac{1}{(-\frac{2}{7})} = \frac{7}{2}$.

Remember $\frac{1}{\left(\frac{a}{b}\right)} = \frac{b}{a}$

$y - (-5) = \frac{7}{2}(x - (-4))$

Using **3**: here $m = \frac{7}{2}$ and $(x_1, y_1) = (-4, -5)$

$y + 5 = \frac{7}{2}(x + 4)$

$2 \times (y + 5) = 2 \times \frac{7}{2}(x + 4)$

Multiply each side by 2 to clear the fraction.

$2(y + 5) = 7(x + 4)$

$2y + 10 = 7x + 28$

$2y = 7x + 18$

$7x - 2y + 18 = 0$

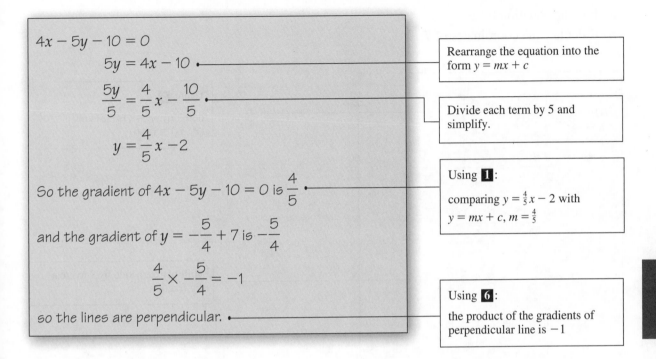

Worked exam style question 2

The line p passes through the points $(1, -3)$ and $(4, 6)$, and the line q has equation $y = \frac{4}{5}x + 5$. The lines p and q intersect at T.

(a) Find an equation for p.

(b) Calculate the coordinates of T.

(a) $$\frac{y-(-3)}{6-(-3)} = \frac{x-1}{4-1}$$

Using **4**: here $(x_1, y_1) = (1, -3)$ and $(x_2, y_2) = (4, 6)$

$$\frac{y+3}{6+3} = \frac{x-1}{3}$$

$$\frac{y+3}{9} = \frac{x-1}{3}$$

$$9 \times \frac{y+3}{9} = 9 \times \frac{x-1}{3}$$

Multiply each side by 9 to clear the fraction.

$$y + 3 = 3(x-1)$$

$$y + 3 = 3x - 3$$

$$y = 3x - 6$$

(b) $$3x - 6 = \frac{4}{5}x + 5$$

To find the point of intersection, solve p and q simultaneously. Substitute the equation for line p, $y = 3x - 6$, into the equation for line q, $y = \frac{4}{5}x + 5$

$$3x = \frac{4}{5}x + 11$$

$$3x - \frac{4}{5}x = 11$$

$$\frac{11}{5}x = 11$$

Divide each side by $\frac{11}{5}$ to clear the fraction, so that $11 \div \frac{11}{5} = 5$

so $$x = 5$$

$$y = 3(5) - 6$$

Work out the y-coordinate. Substitute $x = 5$ into $y = 3x - 6$

$$= 15 - 6$$

$$= 9$$

so the coordinates of T are $(5, 9)$.

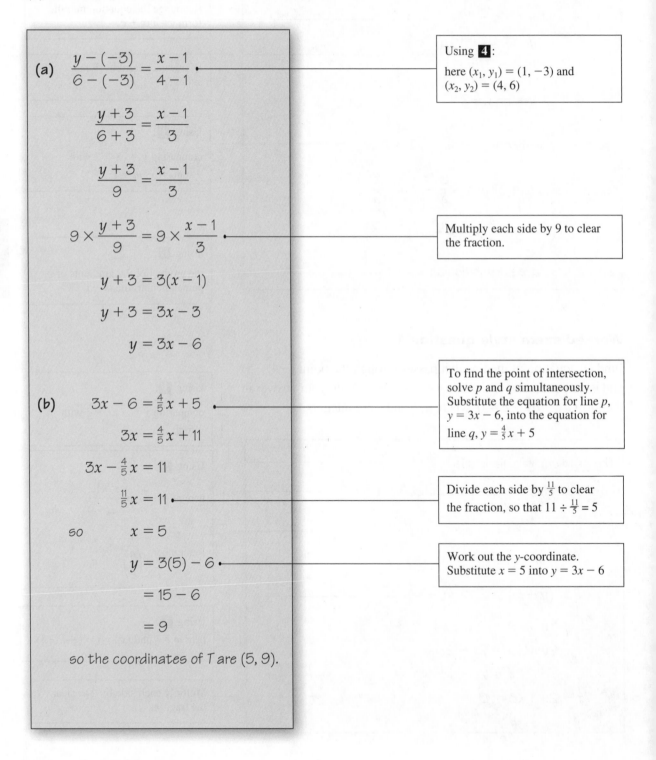

Worked exam style question 3

The points $P(3, -2)$, $Q(1, 2)$ and $R(9, r)$ are the vertices of a triangle, where $\angle PQR = 90°$.

(a) Find the gradient of the line PQ.

(b) Calculate the value of r.

(a) $\frac{2-(-2)}{1-3} = \frac{2+2}{1-3}$

Using **2**: here $(x_1, y_1) = (3, -2)$ and $(x_2, y_2) = (1, 2)$

$= \frac{4}{-2}$

$= -2$

(b) The gradient of $PQ = -2$, so the gradient of $QR = \frac{1}{2}$

Using **5**: here $m = -2$, so that $\frac{-1}{(-2)} = \frac{1}{2}$

$\frac{r-2}{9-1} = \frac{1}{2}$

Form an equation in r using **2**. Here $(x_1, y_1) = (1, 2)$ and $(x_2, y_2) = (9, r)$

$\frac{r-2}{8} = \frac{1}{2}$

$8 \times \frac{r-2}{8} = 8 \times \frac{1}{2}$

Multiply each side by 8 to clear the fraction.

$r - 2 = 4$

$r = 6$

Revision exercise 5

1 A line has gradient $\frac{1}{5}$ and passes through the point $(-4, 2)$. Find the equation of the line in the form $ax + by + c = 0$, where a, b and c are integers.

2 Find an equation of the line that passes through the point $(-4, -1)$ and is perpendicular to the line $y = \frac{1}{4}x + 5$. Write your answer in the form $y = mx + c$.

3 **(a)** The line $y = -\frac{2}{3}x + c$ passes through the point $(4, -3)$. Find the exact value of c.

(b) Show that the line $y = -\frac{2}{3}x + c$ is perpendicular to the line $12x - 8y + 5 = 0$.

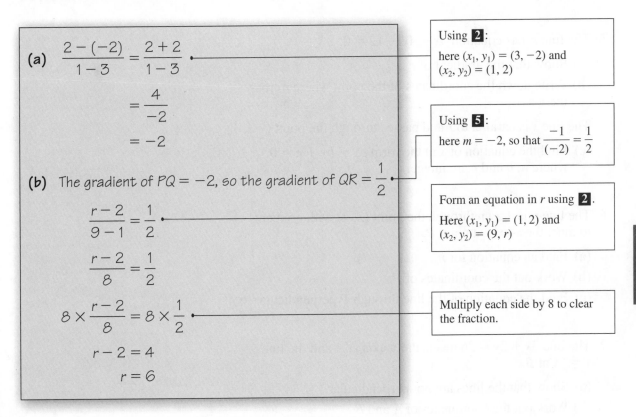

4 A line passes through the points $(-6, 6)$ and $(6, 2)$.

(a) Find the equation of the line in the form $y = mx + c$.

The line crosses the x-axis at point A and the y-axis at point B.

(b) Work out the area of $\triangle OAB$, where O is the origin.

5 The line r has equation $3x - 4y - 12 = 0$.

(a) Show that the gradient of r is $\frac{3}{4}$.

(b) Write down the coordinates of the point where r crosses the y-axis.

The line s is parallel to r and passes through the point $(-2, 1)$.

(c) Find the equation of s in the form $ax + by + c = 0$, where a, b and c are integers.

6 The line n is drawn through the point $(3, 4)$ with gradient $-\frac{2}{3}$ to meet the x-axis at point P.

(a) Find an equation for n.

(b) Work out the coordinates of P.

(c) Find the equation of the line through P perpendicular to n.

7 The line $3x + 2y = 36$ meets the x-axis at A and the line $y = \frac{3}{2}x$ at B.

(a) Show that the lines are not perpendicular.

(b) Work out the coordinates of A and B.

(c) Find the area of $\triangle OAB$, where O is the origin. **E**

8 The points P and Q have coordinates $(2, 6)$ and $(6, 4)$ respectively.

(a) Find the gradient of the line PQ.

A line l is drawn through Q perpendicular to PQ to meet the x-axis at R.

(b) Find an equation of l.

(c) Work out the coordinates of R.

9 The line l_1 passes through the points $(0, -4)$ and $(5, 6)$, and the line l_2 has equation $3x + y = 4$.

(a) Find the equation of l_1 in the form $y = mx + c$.

The lines l_1 and l_2 intersect at Q.

(b) Calculate the coordinates of Q.

10 The line l has equation $5x - 4y + 12 = 0$. The line m passes through the point $(1, -6)$ and is perpendicular to l.

(a) Find an equation of m.

(b) Show that the lines l and m intersect at the point $(-4, -2)$.

11 A line with gradient $\frac{5}{6}$ passes through the points $(5, 2)$ and $(-1, n)$.

(a) Find an equation of the line in terms of x and y only.

(b) Work out the value of n. (E)

12 A line passes through the points with coordinates $(-2, 4)$ and $(4, k)$.

(a) Find the value of k if the line is
(i) parallel to and
(ii) perpendicular to the line $3x - y - 6 = 0$.

(b) Given that $k = 10$, find the equation of the line in the form $y = mx + c$. (E)

13 The points $A(3, 0)$, $B(-3, 2)$ and $C(-2, k)$ are the vertices of a triangle, where $\angle ABC = 90°$.

(a) Find the gradient of the line AB.

(b) Calculate the value of k.

(c) Find an equation of the line that passes through A and C. (E)

14 A line passes through the points with coordinates $(k, 2k - 3)$ and $(2 - k, k - 2)$, where k is a constant.

(a) Show that the gradient of the line is $\frac{1}{2}$.

(b) Find, in terms of k, an equation of the line. (E)

15 The line l_1 has equation $4y = x$.
The line l_2 has equation $y = 5 - 4x$.

(a) Show that these lines are perpendicular.

(b) On the same axes, sketch the graphs of l_1 and l_2. Show clearly the coordinates of all points where the graphs meet the coordinate axes.

The lines l_1 and l_2 intersect at A.

(c) Calculate, as exact fractions, the coordinates of A. (E)

Test yourself	What to review
	If your answer is incorrect
1 Write $y = \frac{1}{3}x - 4$ in the form $ax + by + c = 0$.	*Review Heinemann Book C1 page 66* *Revise for C1 page 35* *Worked exam style question 1*
2 Work out the coordinates of the point where the line $2x - 5y - 3 = 0$ meets the x-axis.	*Review Heinemann Book C1 page 67* *Revise for C1 page 34 Example 1*
3 Work out the gradient of the line joining the points $(a, -2a)$ and $(4a, 4a)$.	*Review Heinemann Book C1 page 68* *Revise for C1 page 37* *Worked exam style question 3*
4 Find an equation of the line that has gradient $\frac{1}{2}$ and passes through the point $(6, -1)$.	*Review Heinemann Book C1 page 70* *Revise for C1 page 34 Example 1*
5 Find an equation of the line that passes through the points $(0, -3)$ and $(7, 0)$.	*Review Heinemann Book C1 pages 72–73* *Revise for C1 page 36* *Worked exam style question 2*
6 Find an equation of the line that passes through the point $(-3, 3)$ and is perpendicular to the line $y = -\frac{1}{2}x + 3$.	*Review Heinemann Book C1 page 77* *Revise for C1 page 35* *Worked exam style question 1*

Test yourself answers

1 $x - 3y - 12 = 0$ **2** $(\frac{3}{2}, 0)$ **3** 2 **4** $y = \frac{1}{2}x - 4$ **5** $y = \frac{3}{7}x - 3$ **6** $y = 2x + 9$

Sequences and series 6

Key points to remember

1 A series of numbers following a set rule is called a sequence. 3, 7, 11, 15, 19, … is an example of sequence.

2 Each number in a sequence is called a **term**. The nth term of a sequence is sometimes called the **general term**.

3 A sequence can be expressed as a formula for the nth term. For example the formula $U_n = 4n + 1$ produces the sequence 5, 9, 13, 17, … by replacing n with 1, 2, 3, 4, etc in $4n + 1$.

4 A sequence can be expressed by a **recurrence relationship**. For example the same sequence 5, 9, 13, 17, … can be formed from $U_{n+1} = U_n + 4$, $U_1 = 5$. (U_1 must be given.)

5 A recurrence relationship of the form

$$U_{k+1} = U_k + n,\ k \geqslant 1 \quad n \in \mathbb{Z}$$

is called an **arithmetic sequence**.

6 All arithmetic sequences can be put in the form

$$a + (a + d) + (a + 2d) + (a + 3d) + (a + 4d) + (a + 5d)$$

↑	↑	↑	↑	↑	↑
1st term	2nd term	3rd term	4th term	5th term	6th term

7 The nth term of an arithmetic series is $a + (n - 1)d$, where a is the first term and d is the common difference.

8 The formula for the sum of an arithmetic series is

$$S_n = \frac{n}{2}[2a + (n - 1)d]$$

or $$S_n = \frac{n}{2}(a + L)$$

where a is the first term, d is the common difference, n is the number of terms and L is the last term in the series.

9 You can use Σ to signify 'sum of'. You can use Σ to write series in a more concise way

e.g. $$\sum_{r=1}^{10}(5 + 2r) = 7 + 9 + \ldots + 25$$

Example 1

The nth term in a sequence is given by $U_n = 4n + 3$

Find (a) the first three terms of the sequence.
(b) the 50th term of the sequence.
(c) the value of n for which $U_n = 87$

(a) $U_1 = 4 \times 1 + 3 = 7$
$U_2 = 4 \times 2 + 3 = 11$
$U_3 = 4 \times 3 + 3 = 15$

Use $U_n = 4n + 3$ with $n = 1, 2$ and 3

(b) $U_{50} = 4 \times 50 + 3$
$= 203$

Use $U_n = 4n + 3$ with $n = 50$

(c) $87 = 4n + 3$
$84 = 4n$
$n = 21$

Replace $U_n = 87$ and solve the equation for n

Example 2

A sequence of terms $\{U_n\}, n \geqslant 1$ is defined by the recurrence relation $U_{n+2} = p\,U_{n+1} - U_n$ where p is a constant. Given also that $U_1 = 1$ and $U_2 = 6$

(a) find an expression in terms of p for U_3
(b) find an expression in terms of p for U_4

Given that $U_4 = 29$

(c) Find possible values of p

(a) $U_3 = p\,U_2 - U_1$ — Substitute $n = 1$ into the recurrence relation
$U_3 = p \times 6 - 1$ — Substitute $U_1 = 1$ and $U_2 = 6$
$U_3 = 6p - 1$

(b) $U_4 = p\,U_3 - U_2$ — Substitute $n = 2$ into the recurrence relation
$= p(6p - 1) - 6$ — Substitute $U_3 = 6p - 1$ and $U_2 = 6$
$= 6p^2 - p - 6$

(c) $29 = 6p^2 - p - 6$ — Set $U_4 = 29$
$0 = 6p^2 - p - 35$ — Subtract 29 from both sides
$0 = (3p + 7)(2p - 5)$ — Factorise to find p
$p = -\frac{7}{3}, \frac{5}{2}$

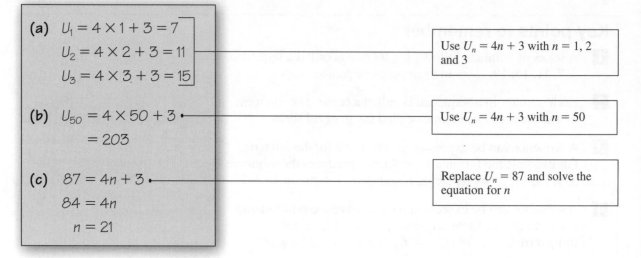

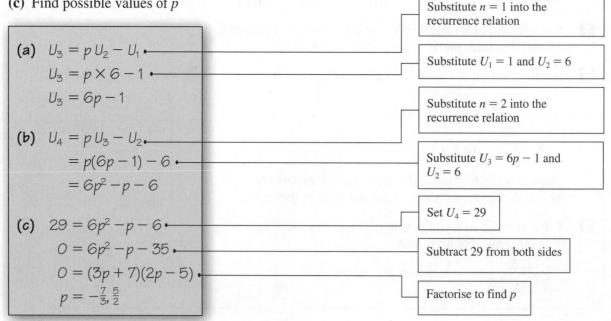

Example 3

For the following arithmetic series

$$4 + 7 + 10 + 13 + 16 + \ldots$$

find **(a)** the 30th term **(b)** the sum to 30 terms.

(a) 30th term $= a + 29d$

$= 4 + 29 \times 3$

$= 4 + 87$

$= 91$

Using **7**:
nth term $= a + (n - 1)d$ with $a = 4$, $d = 3$ and $n = 30$

(b) $S_{30} = \frac{n}{2}\{2a + (n - 1)d\}$

$= \frac{30}{2}\{2 \times 4 + 29 \times 3\}$

$= 15\{8 + 87\}$

$= 15 \times 95$

$= 1425$

Using **8**:
With $S_n = \frac{n}{2}\{2a + (n - 1)d\}$
with $a = 4$, $d = 3$ and $n = 30$

Worked exam style question 1

Given that the 2nd term of an arithmetic series is 8 and the 6th term is -12

(a) find the first term **(b)** find the sum to 10 terms.

2nd term $= 8$, so $a + d = 8$ ①

6th term $= -12$, so $a + 5d = -12$ ②

Using **7**:
nth term $= a + (n - 1)d$ with $n = 2$ and $n = 6$

(a) ② − ① $4d = -20$

$d = -5$

Solve the two equations simultaneously

The common difference is -5

$a + (-5) = 8$

$a = 13$

Substitute $a = -5$ in ①

So the first term is 13

(b) $S_n = \frac{n}{2}\{2a + (n - 1)d\}$

$S_{10} = \frac{10}{2}\{2 \times 13 + 9 \times -5\}$

$= 5\{26 - 45\}$

$= 5 \times -19$

$= -95$

Using **8**:
$S_n = \frac{n}{2}\{2a + (n - 1)d\}$
with $a = 13$, $d = -5$ and $n = 10$

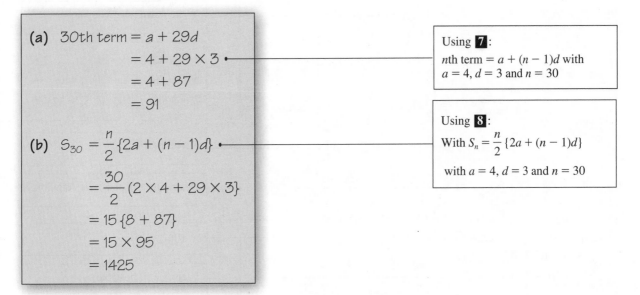

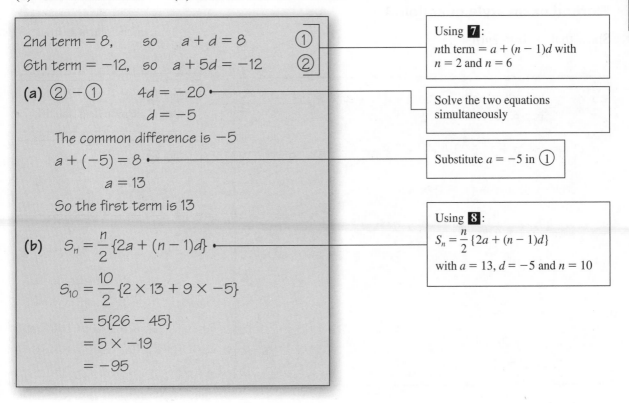

Worked exam style question 2

Each year, Peter pays into a savings scheme. In the first year he pays in £500. His payments then increase by £200 each year, so that he pays £700 in the second year, £900 in the third year and so on. How many years will it be before he has saved £32 000?

$500 + 700 + 900 + \ldots = 32\,000$ — Write down what you know

so $a = 500$, $d = 200$, $S_n = 32\,000$

$$32\,000 = \frac{n}{2}\{2 \times 500 + (n-1) \times 200\}$$

Using **8**: $S_n = \frac{n}{2}\{2a + (n-1)d\}$ with $a = 500$, $d = 200$ and $S_n = 32\,000$

$$32\,000 = \frac{n}{2}\{1000 + 200n - 200\}$$

$$32\,000 = \frac{n}{2}\{800 + 200n\}$$

Solve for n

$$32\,000 = \frac{800 \times n}{2} + \frac{200n \times n}{2}$$

$$32\,000 = 400n + 100n^2$$

Rearrange as a quadratic equation

$$100n^2 + 400n - 32\,000 = 0$$

$\div 100$

$$n^2 + 4n - 320 = 0$$

$$(n - 16)(n + 20) = 0$$

Factorise (if possible)

so $n = 16$ or -20

Accept $n = 16$ as it is a number of years

So it will take Peter 16 years to save £32 000.

Worked exam style question 3

Show that $\sum_{r=1}^{n} 4r = 2n(n + 1)$

$\sum_{r=1}^{n} 4r$ means the sum of $4r$ from $r = 1$ to $r = n$

$$\sum_{r=1}^{n} 4r = 4 \times 1 + 4 \times 2 + 4 \times 3 + \ldots + 4 \times n$$

Using **9**: write out the series and simplify

$$= 4 + 8 + 12 + \ldots + 4n$$

$$= \frac{n}{2}\{2a + (n-1)d\}$$

Using **8**: $S_n = \frac{n}{2}\{2a + (n-1)d\}$ with $a = 4$, $d = 4$ and $n = n$

$$= \frac{n}{2}\{2 \times 4 + (n-1) \times 4\}$$

$$= \frac{n}{2}\{8 + 4n - 4\}$$

$$= \frac{n}{2}\{4n + 4\}$$

$$= \frac{4n^2}{2} + \frac{4n}{2}$$

$$= 2n^2 + 2n$$

$$= 2n(n + 1)$$

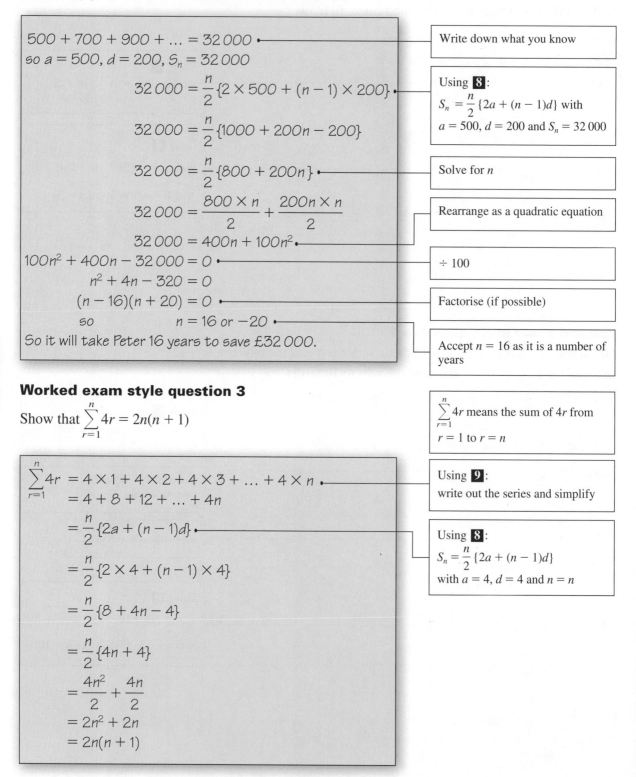

Revision exercise 6

1 The kth term in a sequence is $3k - 4$. Find the first three terms in the sequence.

2 The nth term in a sequence is $2n^2 - 5$. Find n for the term that has a value of 333.

3 A sequence of terms $\{U_n\}$, defined for $n \geqslant 1$, has the recurrence relation $U_{n+2} = kU_{n+1} - 3U_n$.
Given that $U_1 = 4$ and $U_2 = 2$

(a) find an expression, in terms of k, for U_3.

(b) Hence find an expression, in terms of k, for U_4.

(c) Given also that $U_4 = 8$, find possible values of k.

4 Find the 20th terms of the following arithmetic series

(a) $-2, 1, 4, 7, 10, \ldots$

(b) $94, 89, 84, 79, 74, \ldots$

5 Find the sum of the following arithmetic series

(a) $6 + 8 + 10 + \ldots$ to 30 terms

(b) $16 + 19 + 22 + \ldots + 79$

6 Find the value of n such that

$$\sum_{r=1}^{n} (6r + 1) = 2464$$

7 Find the sum of the multiples of 4 less than 100. Hence or otherwise find the sum of the numbers less than 100 which are not multiples of 4.

8 Given that the fourth term of an arithmetic series is 16 and the tenth term is 13, find a and d.

9 In an arithmetic series the fifth term is 0 and the sum to three terms is 9. Find

(a) the first term

(b) the sum to ten terms.

10 A polygon has six sides. The lengths of its sides, starting with the smallest, form an arithmetic series. The perimeter of the polygon is 120 cm and the length of the longest side is three times that of the shortest side. Find for this series

(a) the common difference

(b) the first term. *E*

11 **(a)** Prove that the sum of the first n terms in an arithmetic series is

$$S_n = \frac{n}{2}\{a + L\}$$

where a = first term and L = last term in the series.

(b) Use this result to find the sum of the first 400 natural numbers. *E*

12 Each year for 30 years, David will pay money into a savings scheme. In the first year he pays in £800. His payments then increase by £60 each year, so that he pays £860 in the second year, £920 in the third year and so on.

(a) Find out how much David will pay in the 30th year.

(b) Find the total amount that David will pay in over the 30 years.

(c) David will retire when his savings reach £65 000. Find out how much longer he will have to work. *E*

13 Each year for 40 years, Anne will pay money into a savings scheme. In the first year she pays £500. Her payments then increase by £50 each year, so that she pays £550 in the second year, £600 in the third year and so on.

(a) Find the total amount that Anne will pay in the 40th year.

(b) Find the total amount that Anne will pay in over the 40 years.

Over the same 40 years, Ali will also pay money into the savings scheme. The first year he pays in £890 and his payments then increase by £d each year.

Given that Ali and Anne will pay in exactly the same amount over the 40 years

(c) find the value of d. *E*

14 **(a)** An arithmetic series has first term a and common difference d. Prove that the sum of the first n terms of the series is

$$\tfrac{1}{2}n[2a + (n - 1)d]$$

A company made a profit of £54 000 in the year 2001. A model for future performance assumes that yearly profits will increase in an arithmetic sequence with common difference £d. This model predicts total profits of £619 200 for the nine years 2001 to 2009 inclusive.

(b) Find the value of d.

(c) Using your value of d, find the predicted profit for the year 2011. *E*

Test yourself

What to review

If your answer is incorrect

1 The rth term in a sequence is $3r - 2$. Find

(a) the 10th term and

(b) the value of r when the rth term is 124.

Review Heinemann Book C1 pages 83–84
Revise for C1 page 42 Example 1

---

2 A sequence of terms $\{U_r\}$ $n \geqslant 1$ is defined by the recurrence relation $U_{n+1} = k\,U_n + 7$ where k is a constant.
Given that $U_1 = 6$,

(a) find an expression for U_2 in terms of k

(b) find an expression for U_3 in terms of k.

Given that U_3 is twice as big as U_1

(c) find values of the constant k.

Review Heinemann Book C1 pages 85–87
Revise for C1 page 42 Example 2

---

3 For the following arithmetic series

$$5 + 9 + 13 + \ldots + 213$$

find

(a) the number of terms

(b) the 30th term in the series

(c) the sum of the series.

Review Heinemann Book C1 pages 90–94
Revise for C1 page 43 Example 3

---

4 The fifth term of an arithmetic series is 16 and the sum of the first four terms is 49.

(a) Use algebra to show that the first term of the series is 10 and calculate the common difference.

Review Heinemann Book C1 page 92
Revise for C1 page 43 Worked exam style question 1

(b) Given that the sum to n terms is greater than 1000, find the least possible value of n.

Review Heinemann Book C1 page 94 Example 15

Test yourself answers

1 (a) 28 **(b)** 42 **2 (a)** $6k + 7$ **(b)** $6k^2 + 7k + 7$ **(c)** $\frac{1}{2}, -\frac{5}{3}$ **3 (a)** 53 **(b)** 121 **(c)** 5777 **4 (a)** $d = 1.5$ **(b)** 31

Differentiation

7

Key points to remember

1 The gradient of a curve $y = f(x)$ at a specific point is equal to the gradient of the tangent to the curve at that point.

2 The gradient can be calculated from the gradient function $f'(x)$

3 If $f(x) = x^n$, then $f'(x) = nx^{n-1}$. You reduce the power by 1 and the original power multiplies the expression.

4 The gradient of a curve can also be represented by $\frac{dy}{dx}$.

5 $\frac{dy}{dx}$ is called the **derivative of y with respect to x** and the process of finding $\frac{dy}{dx}$ when y is given is called **differentiation**.

6 If $y = f(x)$, $\frac{dy}{dx} = f'(x)$

7 If $y = x^n$, $\frac{dy}{dx} = nx^{n-1}$ for all real values of n

8 It can also be shown that if $y = ax^n$, where a is a constant then

$$\frac{dy}{dx} = nax^{n-1}$$

You again reduce the power by 1 and the original power multiplies the expression.

9 $y = f(x) \pm g(x)$ then $\frac{dy}{dx} = f'(x) \pm g'(x)$

10 A second order derivative is written as $\frac{d^2y}{dx^2}$ or $f''(x)$, using function notation.

11 You find the rate of change of a function f at a particular point by using $f'(x)$ and substituting in the value of x.

12 The **equation of the tangent** to the curve $y = f(x)$ at point $A(a, f(a))$ is

$$y - f(a) = f'(a)(x - a)$$

13 The **equation of the normal** to the curve $y = f(x)$ at point $A(a, f(a))$ is

$$y - f(a) = -\frac{1}{f'(a)}(x - a)$$

Example 1

Differentiate y with respect to x:

(a) $y = 3x^2$ **(b)** $y = \frac{1}{\sqrt[5]{x}}$ **(c)** $y = 7x^3 + 6x^2 - 1$

(a) $y = 3x^2$

$\frac{dy}{dx} = 2 \times 3x^1$ — Using **7**

so $\frac{dy}{dx} = 6x$

(b) $y = \frac{1}{\sqrt[5]{x}}$

so $y = x^{-\frac{1}{5}}$ — Rewrite using the rules of indices

$\frac{dy}{dx} = \frac{-1}{5} \times x^{-\frac{6}{5}}$ — Using **7**: the power $\frac{-1}{5}$ is reduced to $\frac{-6}{5}$ and the $\frac{-1}{5}$ multiplies the answer

$= \frac{-1}{5(\sqrt[5]{x})^6}$ — You can rewrite the expression

(c) $y = 7x^3 + 6x^2 - 1$

so $\frac{dy}{dx} = 3 \times 7x^2 + 2 \times 6x^1$ — Using **7**

$= 21x^2 + 12x$

Example 2

A curve has equation $y = 5x^2 + 12x + \frac{1}{x^3}$, where $x > 0$.

Find the gradient of the curve at the point (1, 18).

Using **1**: the gradient of the curve is $\frac{dy}{dx}$

$y = 5x^2 + 12x + x^{-3}$ — Express $\frac{1}{x^3}$ as x^{-3} to differentiate

$\frac{dy}{dx} = 10x + 12 - \frac{3}{x^4}$ — Using **8**: $5x^2$ becomes $5 \times 2 \times x^{2-1} = 10x$; $12x$ becomes $1 \times 12 \times x^0 = 12$; x^{-3} becomes $-3x^{-4} = \frac{-3}{x^4}$ as $x^n \to nx^{n-1}$

When $x = 1$, $\frac{dy}{dx} = 10 + 12 - 3$ — Substitute $x = 1$ into expression for $\frac{dy}{dx}$

$\Rightarrow \frac{dy}{dx} = 19$

The gradient of the curve at (1, 18) is 19

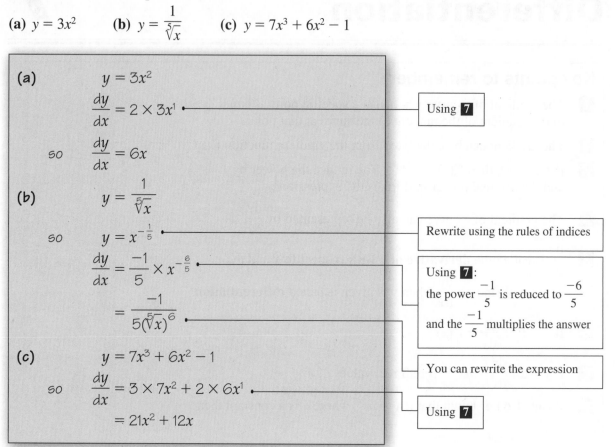

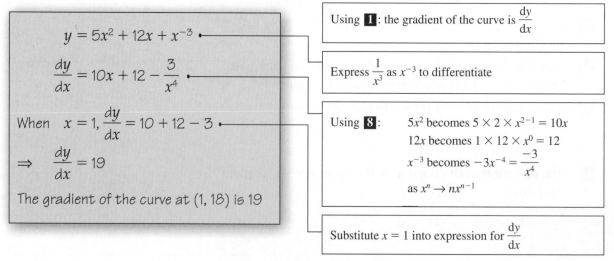

Example 3

Calculate the x-coordinates of the points on the curve with equation $y = x^3 + 4x^2 + 1$, at which the gradient is equal to 16.

$$\frac{dy}{dx} = 3x^2 + 8x$$

Using **2**: find $\frac{dy}{dx}$, the gradient function

Using **1**: differentiate $y = x^3 + 4x^2 + 1$

$\Rightarrow \quad 3x^2 + 8x = 16$

Put gradient function equal to 16

$\Rightarrow \quad 3x^2 + 8x - 16 = 0$

Form a quadratic and solve

$\Rightarrow \quad (3x - 4)(x + 4) = 0$

Factorise or use the formula to find values for x

$\Rightarrow \quad 3x = 4$ or $x = -4$

$\Rightarrow \quad x = \frac{4}{3}$ or $x = -4$

Example 4

(a) Expand $(x + 1)(x - 2)^2$

A curve has equation $y = (x + 1)(x - 2)^2$

(b) Find **(i)** $\frac{dy}{dx}$ and

(ii) $\frac{d^2y}{dx^2}$

(c) Use the relevant expression found in **(b)** to calculate the gradient of this curve at the point with coordinates (1, 2).

(a) $(x + 1)(x^2 - 4x + 4)$

Expand brackets before differentiation

$= x^3 - 4x^2 + 4x + x^2 - 4x + 4$

$= x^3 - 3x^2 + 4$

(b) (i) $\frac{dy}{dx} = 3x^2 - 6x$

Using **8** and **9**

(ii) $\frac{d^2y}{dx^2} = 6x - 6$

Using **8** and **10**: differentiate $\frac{dy}{dx}$ to give $\frac{d^2y}{dx^2}$

(c) $3 - 6 = -3$

So the gradient of the curve at (1, 2) is -3

Using **4**: gradient at the point (1, 2) is given by substituting $x = 1$ into $3x^2 - 6x$

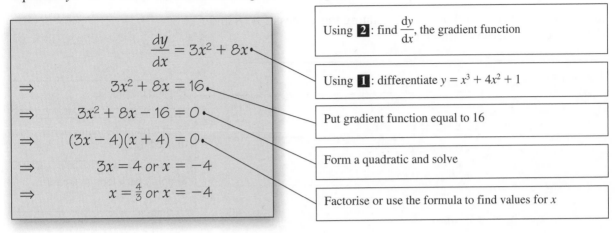

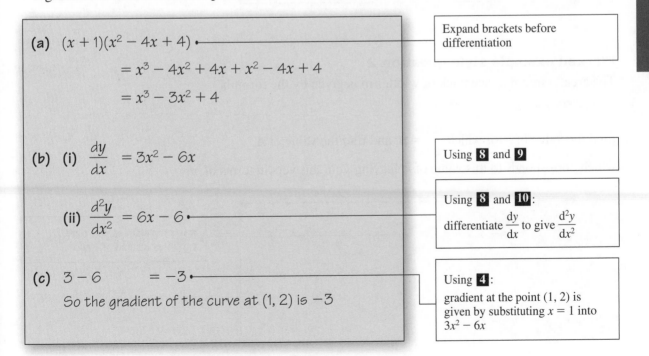

Worked example style question 1

$$f(x) = \frac{(3x^2 + 1)^2}{3x}, x \neq 0$$

(a) Show that $f(x) \equiv Ax + B + Cx^{-1}$, stating the values of A, B and C.
(b) Hence, or otherwise, find $f'(x)$.

(a) $\frac{(3x^2+1)^2}{3x} = \frac{9x^4 + 6x^2 + 1}{3x}$

Expand the bracket

$= \frac{9x^4}{3x} + \frac{6x^2}{3x} + \frac{1}{3x}$

Each term in the numerator is divided by $3x$

$= 3x^3 + 2x + \frac{1}{3}x^{-1}$

Simplify each term to the form ax^n

So $A = 3$, $B = 2$ and $C = \frac{1}{3}$

(b) $\Rightarrow f'(x) = 9x^2 + 2 - \frac{1}{3}x^{-2}$

Using **8** and **9**: this answer is usually sufficient and gains full marks

$= \frac{27x^4 + 6x^2 - 1}{3x^2}$

$= \frac{(9x^2 - 1)(3x^2 + 1)}{3x^2}$

$= \frac{(3x - 1)(3x + 1)(3x^2 + 1)}{3x^2}$

This answer would be useful if the question asked for the solution of $f'(x) = 0$

Worked example style question 2

The area, A m^2, of a race track of width r m is given by the formula

$$A = 500r - \pi r^2$$

Find the value of r for which $\frac{dA}{dr} = 0$, and find the value of A, which corresponds to this value of r, leaving your answers in terms of π.

$\frac{dA}{dr} = 500 - 2\pi r$

π is a constant. A and r are variables instead of y and x

when $\frac{dA}{dr} = 0$

$500 - 2\pi r = 0$

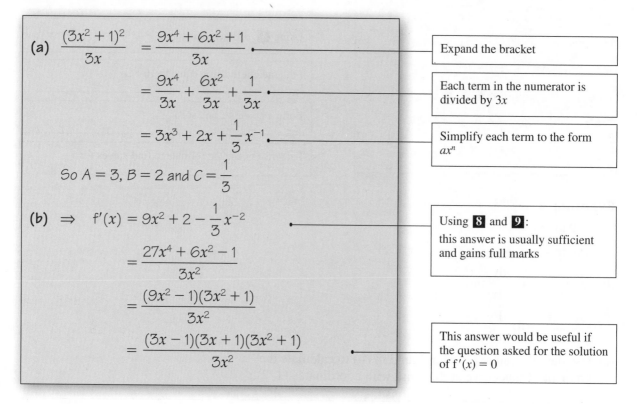

$$\Rightarrow \quad r = \frac{250}{\pi}$$

$$A = 500 \times \frac{250}{\pi} - \pi \times \left(\frac{250}{\pi}\right)^2$$

Substitute $r = \frac{250}{\pi}$ into the expression for A.

$$\Rightarrow \quad A = \frac{62\,500}{\pi}$$

Worked example style question 3

Find an equation of the tangent and of the normal at the point (4, −1) to the curve with equation $y = \frac{16}{x^2} - \sqrt{x},\ x > 0$.

$$y = 16x^{-2} - x^{\frac{1}{2}}$$

Express $\frac{16}{x^2}$ as $16x^{-2}$ and $\sqrt{x}$ as $x^{\frac{1}{2}}$

$$\therefore \quad \frac{dy}{dx} = -32x^{-3} - \tfrac{1}{2}x^{-\frac{1}{2}}$$

Using **8** and **9**

At the point (4, −1), $x = 4$

$$\Rightarrow \quad \frac{dy}{dx} = -32(4)^{-3} - \tfrac{1}{2}(4)^{-\frac{1}{2}}$$

$$= -\tfrac{32}{64} - \tfrac{1}{4}$$

$$= -\tfrac{3}{4}$$

This is the gradient of the curve, which is also the gradient of the tangent.

The equation of the tangent is

$$y - (-1) = -\tfrac{3}{4}(x - 4)$$

Using **12**

$$\Rightarrow \quad y + 1 = -\tfrac{3}{4}x + 3$$

$$\Rightarrow \quad y = -\tfrac{3}{4}x + 2$$

The gradient of the normal is $+\frac{4}{3}$

The normal is perpendicular to the tangent. The normal gradient $= \frac{-1}{\text{gradient of tangent}}$

Then the equation of the normal is

$$y - (-1) = \tfrac{4}{3}(x - 4)$$

Using **13**

$$\Rightarrow \quad 3y + 3 = 4x - 16$$

$$\Rightarrow \quad 3y = 4x - 19$$

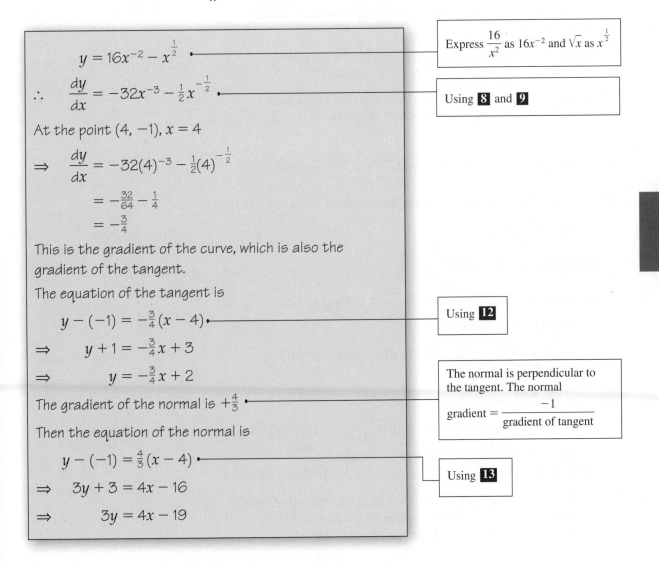

Revision exercise 7

1 $y = x^3 - 7x^2 + 15x + 3, x \geqslant 0.$

Find $\frac{dy}{dx}$

2 Differentiate with respect to x

$$3x^2 - 5\sqrt{x} + \frac{1}{2x^2}$$

3 Differentiate with respect to x

$$2x^4 + \frac{x^2 + 2x}{\sqrt{x}}$$

4 Find the gradient of the curve C, with equation $y = 5x^2 + 4x - 4$, at each of the points $(1, 5)$ and $(-2, 8)$.

5 Find the gradient of the curve C, with equation $y = x^3 - 8x^2 + 16x + 2$, at each of the points where C meets the line $y = x + 2$

6 Find the coordinates of the two points on the curve with equation $y = x^3 - 6x + 2$ where the gradient is 6.

7 Find the coordinates of the two points on the curve with equation $y = x^3 + 4x^2 - x$ where the tangents are parallel to the line $y = 2x$

8 Given that $x = 8 + 12t - 6t^2$, find $\frac{dx}{dt}$ and $\frac{d^2x}{dt^2}$ in terms of t.

9 The area of a surface is related to the radius of its base by the formula $A = \pi r^2 + \frac{729}{r}$

(a) Find an expression for $\frac{dA}{dr}$

(b) Find the value of r for which $\frac{dA}{dr} = 0$, leaving your answer in terms of π.

10 The volume, $V\,\text{cm}^3$, of a solid of radius r cm is given by the formula

$$V = \pi(300r - r^3)$$

Find the positive value of r for which $\frac{dV}{dr} = 0$, and find the value of V, which corresponds to this value of r.
(Take π as 3.142 and give your answer to 3 significant figures.)

11 The line l is the tangent, at the point $(2, 3)$, to the curve with equation $y = a + bx^2$, where a and b are constants. The tangent l has gradient 8. Find the values of a and b.

12 A curve has equation $y = x^{\frac{3}{2}} + 48x^{-\frac{1}{2}}, x > 0$

(a) Show that $\frac{dy}{dx} = Ax^{-\frac{3}{2}}(x^n - B)$, where A, n and B are constants and find the values of these constants.

(b) Find the coordinates of the point on the curve where the gradient is zero.

(c) Find the equation of the tangent to this curve at the point (1, 49).

13 The function f is defined by $f(x) = \frac{x^2}{2} + \frac{8}{x^2}, x \in \mathbb{R}, x \neq 0$

(a) Find $f'(x)$ **(b)** Solve $f'(x) = 0$

14 $f(x) = \frac{(x^2 - 3)^2}{x^3}, x \neq 0$

(a) Show that $f(x) \equiv x - 6x^{-1} + 9x^{-3}$

(b) Hence, or otherwise, differentiate $f(x)$ with respect to x.

(c) Find the values of x for which $f'(x) = 0$. **E**

15 The figure shows part of the curve C with equation $y = f(x)$ where $f(x) = x^3 - 6x^2 + 5x$

The curve crosses the x-axis at the origin O and at the points A and B.

(a) Factorise $f(x)$ completely.

(b) Write down the x-coordinates of the points A and B.

(c) Find the gradient of C at A. **E**

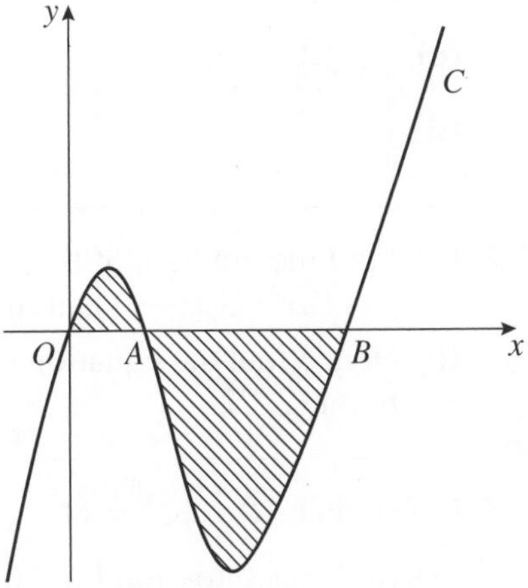

16 The figure shows part of the curve C with equation $y = \frac{3}{2}x^2 - \frac{1}{4}x^3$

The curve C touches the x-axis at the origin and passes through the point $A(p, 0)$.

(a) Show that $p = 6$

(b) Find an equation of the tangent to C at A. **E**

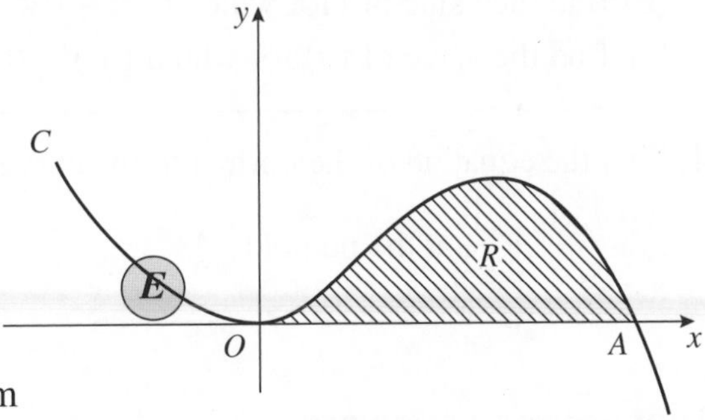

17 **(a)** Show that $\frac{(3x - 1)^2}{x^2}$ may be written in the form $L + \frac{M}{x} + \frac{N}{x^2}$, giving the values of L, M and N.

(b) Hence find the gradient at the point (1, 4) of the curve C with equation $y = \frac{(3x - 1)^2}{x^2}$

(c) Find the equation of the normal to C at (1, 4)

18 For the curve C with equation $y = \text{f}(x)$, $\dfrac{\text{d}y}{\text{d}x} = 3x^2 - 4x - \dfrac{2}{3x}$

(a) Find $\dfrac{\text{d}^2y}{\text{d}x^2}$

(b) Find the equation of the normal to the curve C at the point $(1, 2)$, which lies on the curve C.

Test yourself

1 Find $\dfrac{\text{d}y}{\text{d}x}$ and $\dfrac{\text{d}^2y}{\text{d}x^2}$ when

(a) $y = 2x^3 - 5x^2 + 4x + 7$

(b) $y = \dfrac{3}{x^2} + 5\sqrt{x}$

(c) $y = (2x - 9)^2$

(d) $y = \dfrac{12x + 7}{x^2}$

What to review — If your answer is incorrect: *Review Heinemann Book C1 pages 113–115. Revise for C1 page 50 Example 1*

---

2 **(a)** Find the gradient of the curve with equation $y = (2x + 3)(x - 1)$, at the point with coordinates $(2, 7)$.

(b) Hence find the equation of the tangent to the curve at the point $(2, 7)$.

Review Heinemann Book C1 pages 117–119. Revise for C1 pages 52–53. Worked exam style question 3

---

3 Given that $\text{f}(t) = 8t^{\frac{1}{2}} + 6t^{-\frac{1}{2}}, t > 0$

(a) find an expression for $\text{f}'(t)$

(b) find the value of t for which $\text{f}'(t) = 0$

(c) find the value of $\text{f}(t)$ for which $\text{f}'(t) = 0$

Review Heinemann Book C1 pages 116–117. Worked exam style question 2

---

4 Find the equation of the normal to the curve

$$y = \frac{1}{x} + \frac{1}{x^2} \text{ at the point } \left(1, \tfrac{1}{2}\right)$$

Review Heinemann Book C1 page 118. Worked exam style question 3

Test yourself answers

1 (a) $6x^2 - 10x + 4$, $12x - 10$ **(b)** $-6x^{-3} + \frac{5}{2}x^{-\frac{1}{2}}$, $18x^{-4} - \frac{5}{4}x^{-\frac{3}{2}}$ **(c)** $8x - 36$, 8 **(d)** $-12x^{-2} - 14x^{-3}$, $24x^{-3} + 42x^{-4}$ **2 (a)** 9, **(b)** $y = 9x - 11$

3 (a) $4t^{-\frac{1}{2}} - 3t^{-\frac{3}{2}}$ **(b)** $t = \frac{3}{4}$ **(c)** $8\sqrt{3}$ **4** $y = \frac{1}{3}x + \frac{1}{6}$

Integration

8

Key points to remember

1 $\int x^n \, dx = \frac{x^{n+1}}{n+1} + c \quad (n \neq -1)$

2 $\int kx^n \, dx = k\frac{x^{n+1}}{n+1} + c \quad (n \neq -1)$

3 $\int (kx^n + lx^m) \, dx = k\frac{x^{n+1}}{n+1} + l\frac{x^{m+1}}{m+1} + c \quad (n \neq -1, m \neq -1)$

4 You can use rules of indices to write the expression in a suitable form to integrate.

(i) $\sqrt{x} = x^{\frac{1}{2}}$ (ii) $\frac{1}{x^m} = x^{-m}$

5 You can use algebra to simplify expressions before integrating.

6 You can find the constant of integration, c, when you are given any point (x, y) that the curve passes through.

Example 1

Find **(a)** $\int (3x^2 + 2x^{-2}) \, dx$ **(b)** $\int (x^{-\frac{1}{2}} + 4) \, dx$

(a) $\int (3x^2 + 2x^{-2}) \, dx = 3\frac{x^3}{3} + 2\frac{x^{-1}}{-1} + c$

Using **3**: integrate term by term and do not forget the $+c$

$= x^3 - 2x^{-1} + c$

Simplify each term

(b) $\int (x^{-\frac{1}{2}} + 4) \, dx = \frac{x^{\frac{1}{2}}}{\frac{1}{2}} + 4x + c$

Using **2** and **3**: remember the integral of 4 is simply $4x$

$= 2x^{\frac{1}{2}} + 4x + c$

Sometimes you have to use the rules of indices to make sure that the expression to be integrated is in the form x^n.

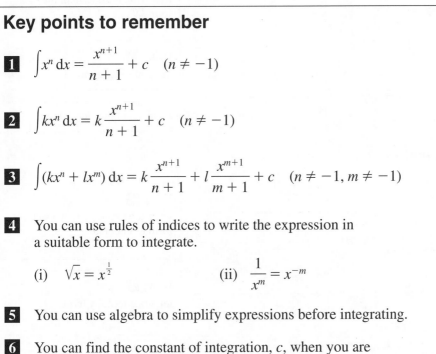

Example 2

Find $\int\left(3\sqrt{x}+\frac{5}{x^3}\right)dx$

$$\int\left(3\sqrt{x}+\frac{5}{x^3}\right)dx = \int(3x^{\frac{1}{2}}+5x^{-3})\,dx$$

Using **4**: first write each term in the form kx^n

$$= 3\frac{x^{\frac{3}{2}}}{\frac{3}{2}}+5\frac{x^{-2}}{-2}+c$$

Using **3**: integrate each term

$$= 2x^{\frac{3}{2}}-\tfrac{5}{2}x^{-2}+c$$

Finally, simplify each term

Worked exam style question 1

Find $\int(x+2\sqrt{x})^2\,dx$

$$\int(x+2\sqrt{x})^2\,dx = \int(x^2+4x^{\frac{3}{2}}+4x)\,dx$$

Using **4**: write the expression in a suitable form to integrate
First square the bracket

$$= \frac{x^3}{3}+4\frac{x^{\frac{5}{2}}}{\frac{5}{2}}+4\frac{x^2}{2}+c$$

Using **3**: integrate term by term

$$= \frac{x^3}{3}+\frac{8}{5}x^{\frac{5}{2}}+2x^2+c$$

Simplify

Worked exam style question 2

The curve C with equation $y = f(x)$ is such that $\frac{dy}{dx} = 3\sqrt{x}+\frac{12}{\sqrt{x}}$, $x > 0$

The curve C passes through the point (4, 30). Using integration, find $f(x)$.

E

$$\int\left(3\sqrt{x}+\frac{12}{\sqrt{x}}\right)dx = \int(3x^{\frac{1}{2}}+12x^{-\frac{1}{2}})\,dx$$

Integrate to find y.
Using **4**: write each term in the form x^n

So $y = 3\frac{x^{\frac{3}{2}}}{\frac{3}{2}}+12\frac{x^{\frac{1}{2}}}{\frac{1}{2}}+c$

So $y = 2x^{\frac{3}{2}}+24x^{\frac{1}{2}}+c$

Using (4, 30), $30 = 2\times2^3+24\times2+c$

Using **6**: find c
Remember that $4^{\frac{3}{2}}$ is 2^3

So $c = 30-16-48 = -34$

and $f(x) = 2x^{\frac{3}{2}}+24x^{\frac{1}{2}}-34$

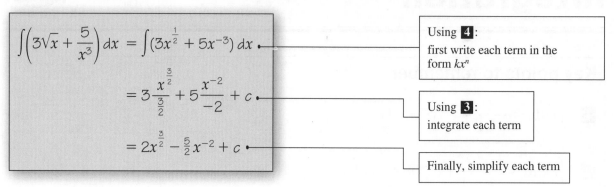

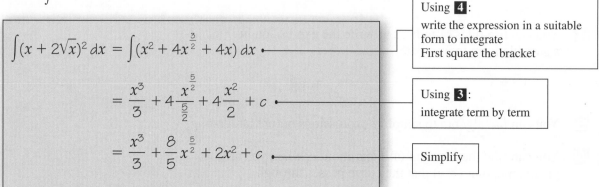

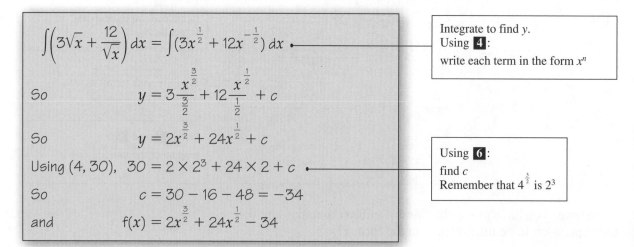

Revision exercise 8

1 Find $\int(\frac{3}{2}x^2 - \frac{1}{4}x^3)\,dx$

2 Find $\int\left(\frac{2}{x^3} - 5x^{\frac{3}{2}}\right)dx$

3 Find $\int\left(\frac{2x^2 - \sqrt{x}}{x}\right)dx$

4 Find $\int\left(\frac{1}{x} - x\right)^2 dx$

5 Find $\int(x-2)(3x+1)\,dx$

6 Find **(a)** $\int(5x + 2\sqrt{x})\,dx$ **(b)** $\int(5x + 2\sqrt{x})^2\,dx$

7 $f(x) = \dfrac{(2x+3)(x-1)}{\sqrt{x}}, x > 0$

(a) Show that f(x) can be written in the form $Px^{\frac{3}{2}} + Qx^{\frac{1}{2}} + Rx^{-\frac{1}{2}}$, stating the values of the constants P, Q and R.

(b) Find $\int f(x)\,dx$

8 $g(x) = \dfrac{(2x+1)(x+1)}{\sqrt{x}}, x > 0$

(a) Show that g(x) can be written in the form $Px^{\frac{3}{2}} + Qx^{\frac{1}{2}} + Rx^{-\frac{1}{2}}$, stating the values of the constants P, Q and R.

(b) Find $\int g(x)\,dx$

9 Find $\int\left(6x^2 - \frac{4}{x^2} + \sqrt[3]{x}\right)dx$

10 Find $\int\left(12x^3 - 3\sqrt{x} + 5\right)dx$

11 Find $\int\left(\sqrt[3]{x} - \frac{1}{\sqrt[3]{x}}\right)^2 dx$

12 The curve C with equation $y = f(x)$ passes through the point (4, 3).

Given that

$$f'(x) = \frac{3x-1}{\sqrt{x}} - 5$$

find f(x). *E*

13 For the curve C with equation $y = \mathrm{f}(x)$

$$\frac{\mathrm{d}y}{\mathrm{d}x} = x^3 + 2x - 6.$$

Given that the point (2, 5) lies on C, find y in terms of x. *(E)*

14 The curve C has equation $y = \mathrm{f}(x)$ and C passes through the origin.
Given that $\mathrm{f}'(x) = 3x^2 - 2x - 6$

(a) find $\mathrm{f}(x)$

(b) sketch C. *(E)*

15 The curve C has equation $y = \mathrm{f}(x)$ and passes through the point (1, 0).
Given that $\mathrm{f}'(x) = (3x - 1)(x - 1)$

(a) find $\mathrm{f}(x)$

(b) sketch C. *(E)*

Test yourself	What to review
	If your answer is incorrect
1 Find $\int (3x^2 - 2x^{-\frac{1}{2}} + 5)\,\mathrm{d}x$	*Review Heinemann Book C1 pages 122–126* *Revise for C1 page 57 Example 1*
2 Find $\int x(5\sqrt{x} - 2)\,\mathrm{d}x$	*Review Heinemann Book C1 pages 126–127* *Revise for C1 page 58 Example 2*
3 Find $\int (3x - 1)(x + 1)\,\mathrm{d}x$	*Review Heinemann Book C1 pages 126–127* *Revise for C1 page 58 Worked exam style question 1*
4 The curve C has equation $y = \mathrm{f}(x)$ and the point (2, 3) lies on C. Given that $\mathrm{f}'(x) = 3x^2 - 6x + 5$ **(a)** find $\mathrm{f}(x)$ **(b)** verify that the point (1, 0) lies on C.	*Review Heinemann Book C1 pages 128–130* *Revise for C1 page 58 Worked exam style question 2*

Test yourself answers

1 $x^3 - 4x^{\frac{1}{2}} + 5x + c$ **2** $2x^{\frac{5}{2}} - x^2 + c$ **3** $x^3 + x^2 - x + c$ **4 (a)** $x^3 - 3x^2 + 5x - 3$

Examination style paper

You may not use a calculator when answering this paper.

You must show sufficient working to make your methods clear.

Answers without working may gain no credit.

1 Solve the inequality

$4(3 - x) > 11 - 5(4 + 2x)$ **(3 marks)**

2 A sequence $U_1, U_2, U_3 \ldots$ is defined by $U_n = n^2 - 3n$.

(a) Find the value of n for which $U_n = 40$ **(2 marks)**

(b) Evaluate $\sum_{r=1}^{6} U_r$ **(3 marks)**

3 On separate diagrams, sketch the curves with equation

(a) $y = x^3$ **(1 mark)**

(b) $y = x^3 + 1$ **(2 marks)**

(c) $y = (x - 2)^3$ **(2 marks)**

On each diagram, show the coordinates of any points at which the curve crosses the coordinate axes.

4 Find in the form $a + b\sqrt{3}$, where a and b are integer constants

(a) $(1 + \sqrt{12})^2$ **(3 marks)**

(b) $\dfrac{2}{4 - \sqrt{12}}$ **(3 marks)**

5 Given that $y = 4x^3 - \dfrac{1}{x^3}$

(a) find $\dfrac{dy}{dx}$ **(3 marks)**

(b) find $\int y\,dx$. **(4 marks)**

6 For the curve C with equation $y = f(x)$

$f'(x) = (3 - \sqrt{x})^2$

Given that C passes through the point $(4, 7)$, find $f(x)$. **(7 marks)**

7 The third and fifth terms of an arithmetic series are 48 and 45 respectively.

(a) Find the first term and the common difference of the series. **(3 marks)**

(b) Find the sum of the first 8 terms of the series. **(2 marks)**

The sum of the first n terms of the series is zero.

(c) Find the value of n. **(3 marks)**

8 For the curve C with equation $y = 2x^3 - 2x + \frac{4}{x}$, $x > 0$

(a) find $\frac{dy}{dx}$ and $\frac{d^2y}{dx^2}$ **(5 marks)**

(b) verify that the gradient of C at $x = 1$ is zero **(2 marks)**

(c) find, in the form $y = mx + c$, the equation of the tangent to C at the point where $x = 2$ **(3 marks)**

9 The straight line l is parallel to the line $2y + 5x = 4$, and passes through the point (3, 1).

(a) Find an equation for l in the form $ax + by + c = 0$, where a, b and c are integer constants. **(4 marks)**

The curve C has equation $y = x^2 + 3x + 2$

(b) Sketch the graph of C, showing the coordinates of the points at which C intersects the axes. **(3 marks)**

(c) Find the coordinates of the points at which l intersects C. **(5 marks)**

10 $f(x) = x^2 + kx + (k + 3)$, where k is a constant. Given that the equation $f(x) = 0$ has equal roots

(a) find the possible values of k **(4 marks)**

(b) solve $f(x) = 0$ for each possible value of k. **(3 marks)**

Given instead that $k = 8$

(c) express $f(x)$ in the form $(x + a)^2 + b$, where a and b are constants **(3 marks)**

(d) solve $f(x) = 0$, giving your answers in surd form. **(2 marks)**

Answers to revision questions

Revision exercise 1

1 (a) $24x^{10}$ **(b)** $16y^5$

2 (a) $(3x + 2y)(3x - 2y)$ **(b)** $3y(2y - 5)$ **(c)** $2(3x + 1)(2x - 3)$

2 (d) $(2x + 3)(x + 1)$ **(e)** $(3 + 2x)(2 - x)$

3 (a) ± 4 **(b)** 16 **(c)** $\frac{1}{49}$ **(d)** $\frac{5}{2}$ **(e)** $\frac{4}{9}$ **(f)** $\frac{9}{64}$ **4** $2\sqrt{5}$

5 (a) ± 4 **(b)** ± 24

6 (a) $\frac{1}{4}(3 + \sqrt{5})$ **(b)** $\dfrac{\sqrt{3} + 1}{2}$ **(c)** $\dfrac{19 + 8\sqrt{5}}{2}$

7 $\sqrt{3}$ **8 (a)** $4\sqrt{7}$ **(b)** $14 - 6\sqrt{5}$ **9 (a)** $k = 3$ **(b)** $x = 3$

10 (a) ± 9 **(b)** 27 **(c)** $\frac{1}{27}$

Revision exercise 2

1 (a) $x = 0$ or $x = -5$ **(b)** $x = -1$ or $x = -6$

1 (c) $x = -\frac{1}{3}$ or $x = \frac{3}{2}$ **(d)** $x = 0$ or $x = \frac{3}{2}$ or $x = 2$

2 (a) $(x + 6)^2 - 36$ **(b)** $5(x - \frac{6}{5})^2 - \frac{36}{5}$ **(c)** $5(x + \frac{3}{2})^2 - \frac{45}{4}$

3 $-\frac{1}{3} \pm \frac{4}{3}$ **4** $-\frac{1}{2} \pm \dfrac{\sqrt{37}}{2}$

5 (a) $x = \dfrac{\pm 2 - 3}{2}$ **(b)** $\dfrac{-7 \pm \sqrt{137}}{4}$ **6** $\frac{1}{2} \pm \frac{1}{2}\sqrt{\frac{17}{5}}$

7 (a)

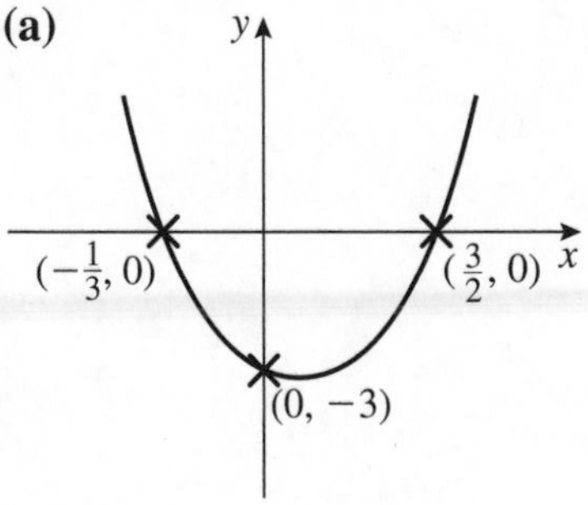

(b)

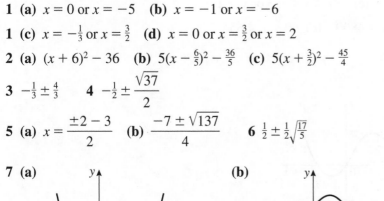

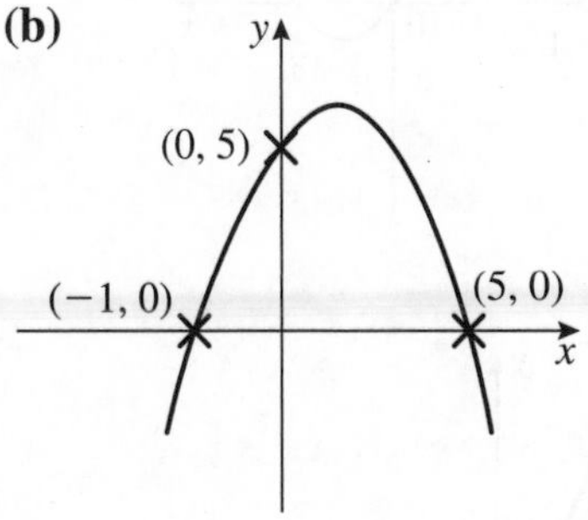

8 $-\frac{1}{2} \pm \sqrt{2}$ **9 (a)** $x = 0$ or -3 **(b)** $c = 9, x = -\frac{3}{2}$

10 (a) $x = -k \pm \sqrt{k^2 + 7}$ **(b)** $k^2 \geqslant -7$ true $\forall k \in \mathbb{R}$

11 $p = 2, q = 5$

12 (a) $x = 0, y = 15$; when $y = 0, x = -5$ or $x = \frac{2}{3}$

12 (b)

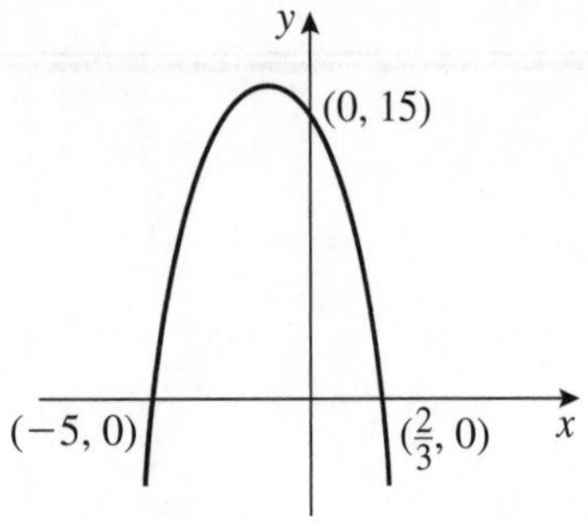

13 $k \geqslant \frac{2}{25}$ or $k \leqslant 0$

14 (a) $p = 3, q = 2, r = -7$ **(b)** minimum value $= -7$

14 (c) $x = -2 \pm \sqrt{\frac{7}{3}}$

Revision exercise 3

1 $x = 4, y = 1\frac{1}{2}$ **2** $(5\frac{1}{2}, -\frac{1}{2})$

3 $(-7, -2)$ and $(14, 5)$ **4** $(3, -2)$ and $(18, 8)$

5 $x > -1\frac{4}{5}$ **6** $x > 1\frac{1}{3}$ **7** $-3 < x < 5$

8 $-3 < x < 12$ **9** $x < -6$ or $x > 2\frac{1}{2}$

10 (a) $x > -7$ **(b)** $x < -3$ or $x > 2$ **(c)** $-7 < x < -3$ or $x > 2$

11 (a) $4x + 40 < 300$ **(b)** $x(x + 20) > 4800$ **(c)** $60 < x < 65$

Revision exercise 4

1 (a) $x(2x - 1)(x + 1)$ **(b)**

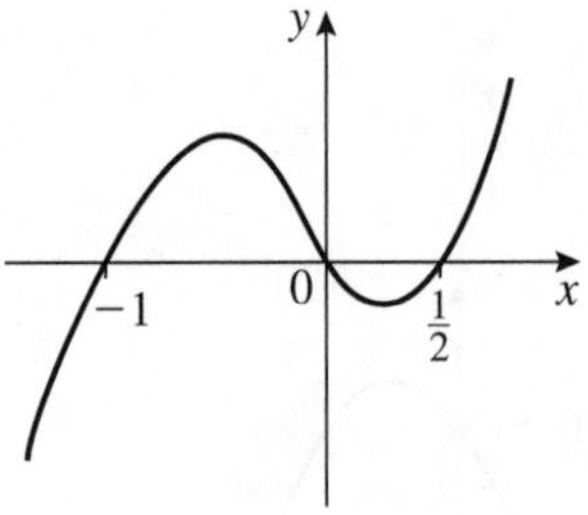

2 (a) $-x(x - 3)(x - 2)$ **(b)**

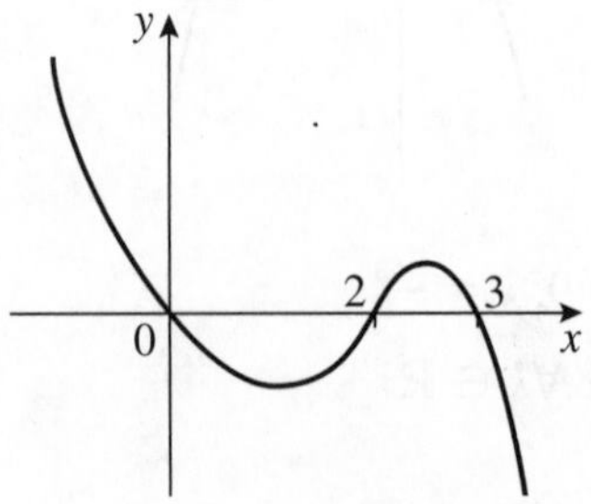

3 (a)

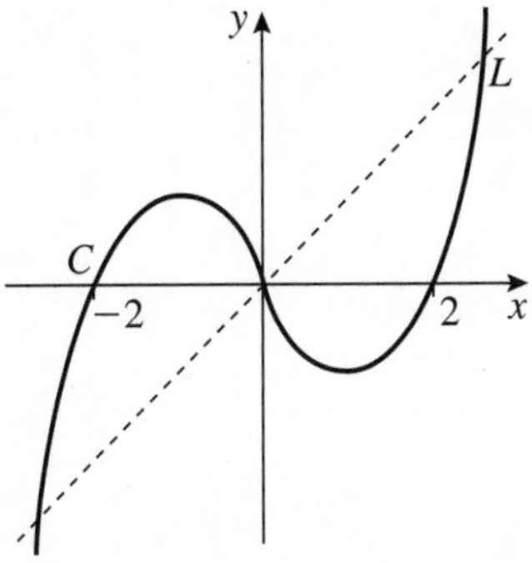

3 (b) (0, 0), (2, 0), (−2, 0) **(c)** (0, 0), (3, 15), (−3, −15)

4 (a)

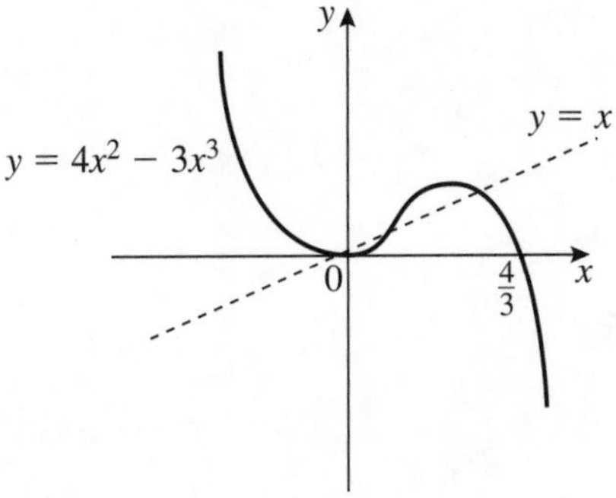

4 (b) (0, 0), $(\frac{1}{3}, \frac{1}{3})$, (1, 1)

5 (a)

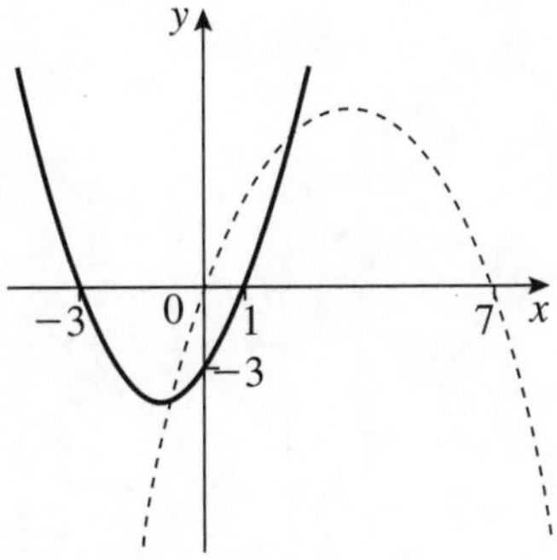

(b) $A(-\frac{1}{2}, -\frac{15}{4})$ $B(3, 12)$

6 (a)

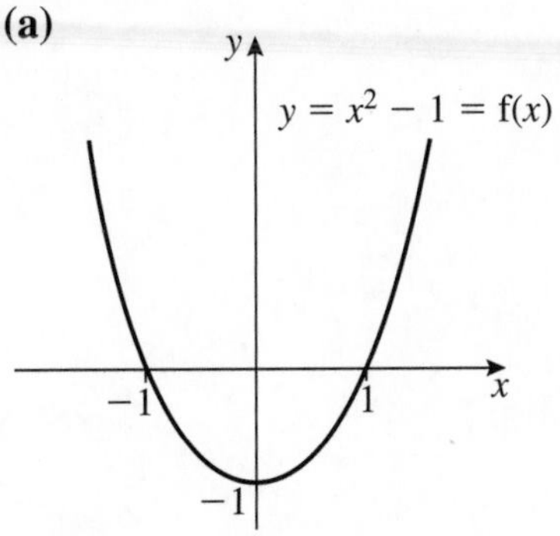

(b)

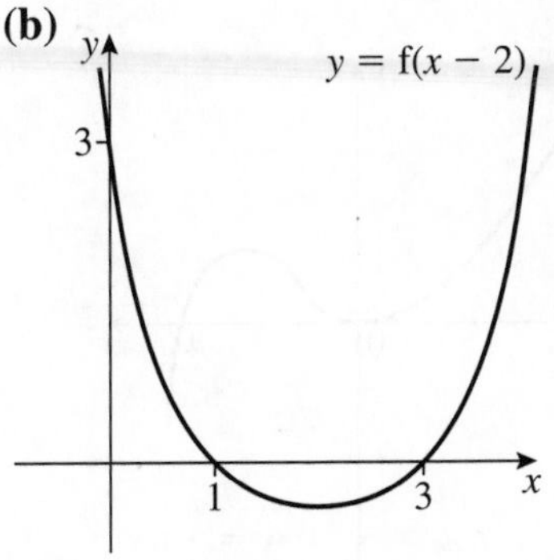

6 (c)

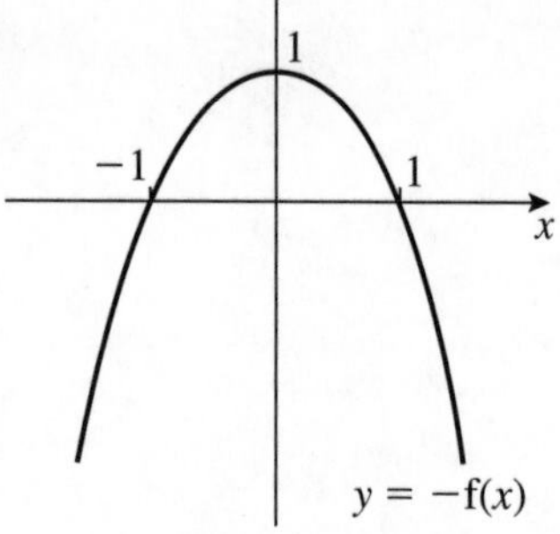

7 (a)

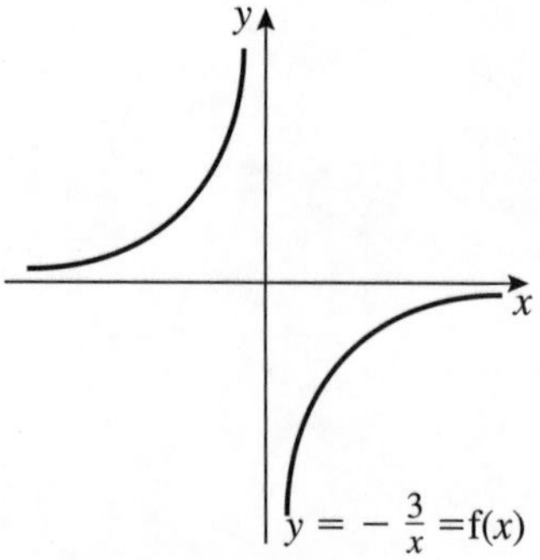

(b)

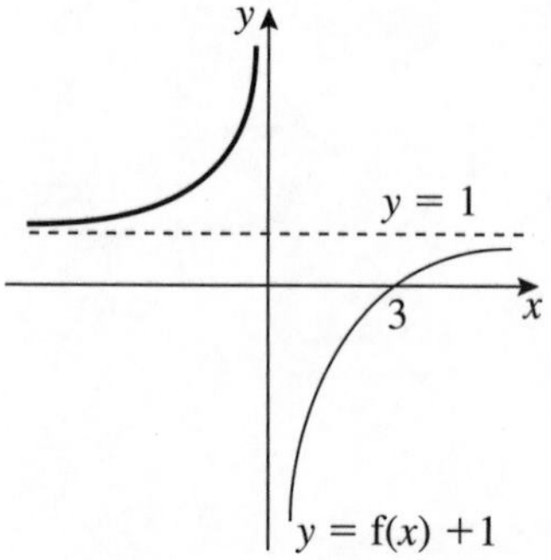

7 (c)

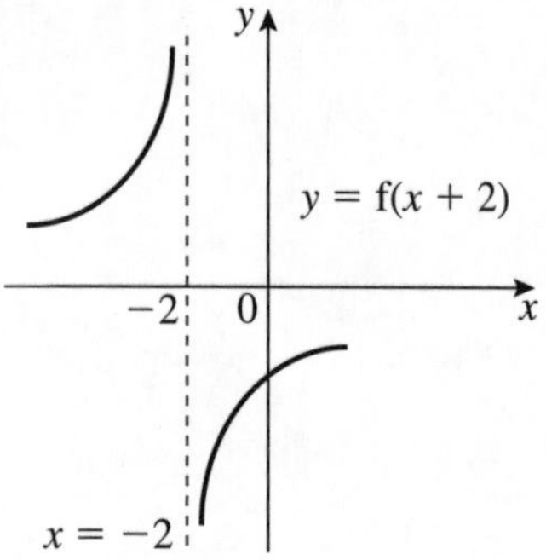

8 (a)

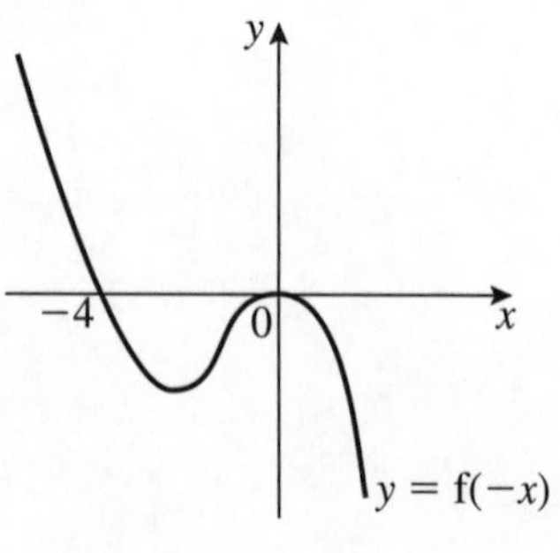

(b)

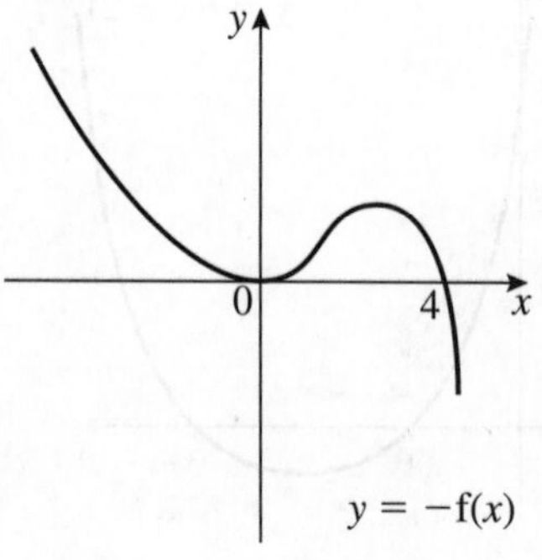

8 (c)

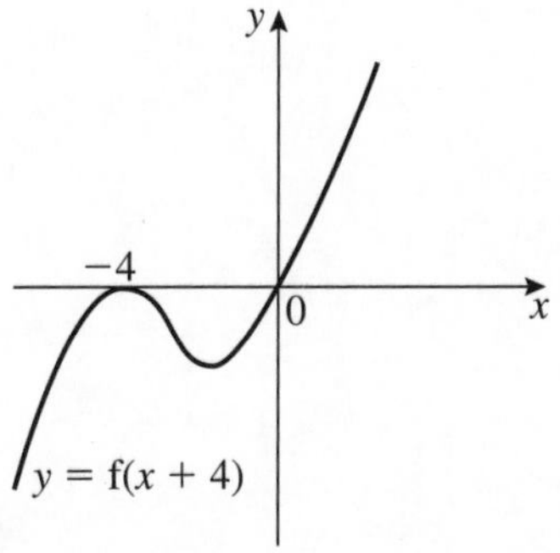

9 (a) $a = -8, b = 15$ **(b)** $k = \frac{1}{4}$

10 (a) $a = 2\frac{1}{2}$ **(b)**

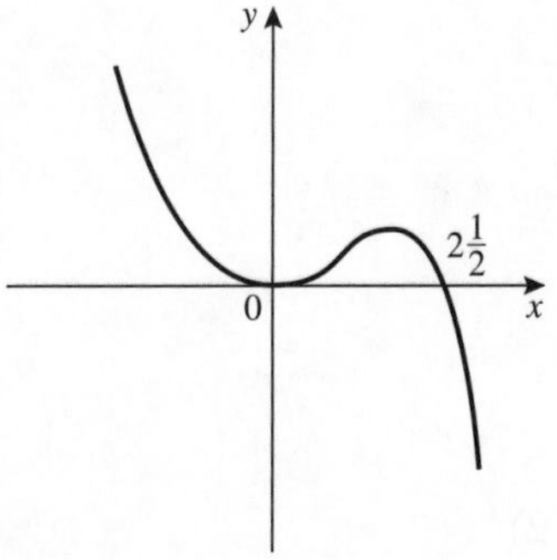

10 (c) $b = 1\frac{1}{2}$ **(d)** $(0, 1\frac{1}{2})$

11

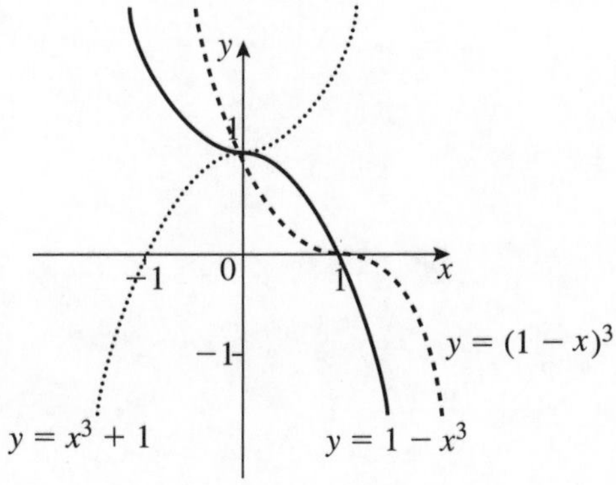

12 (a)

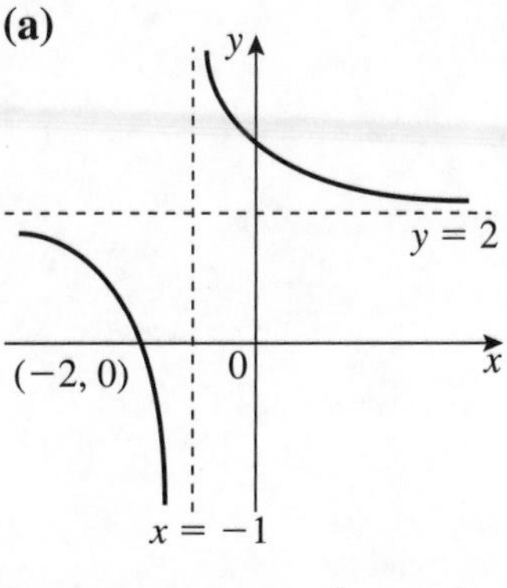

Asymptote, $x = -1$, $y = 2$

(b)

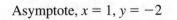

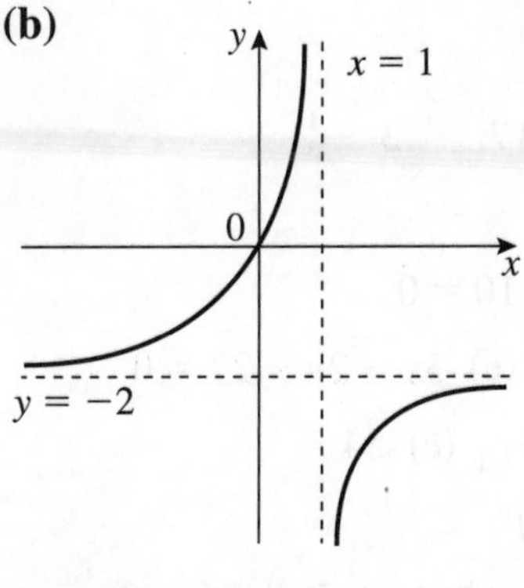

Asymptote, $x = 1$, $y = -2$

12 (c)

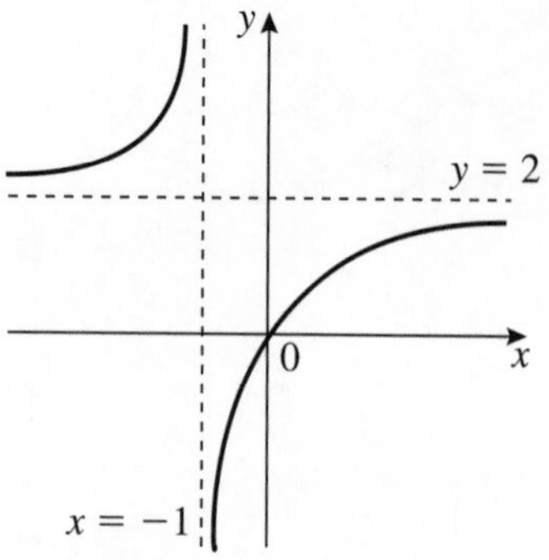

Asymptote, $x = -1, y = 2$

13 (a)

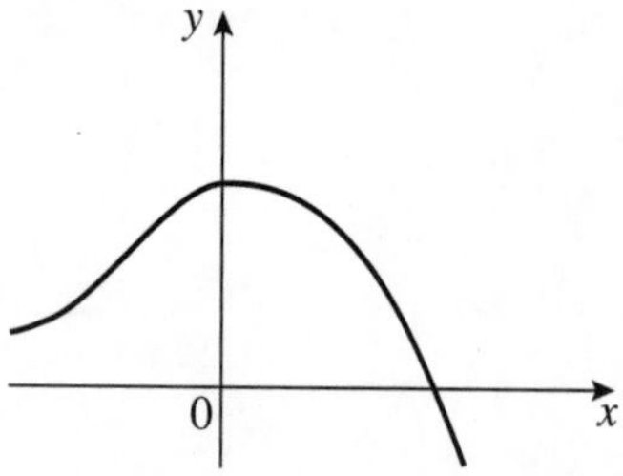

maximum where curve crosses the y-axis at (0, 2)
curve croses the x-axis at (2, 0)

13 (b)

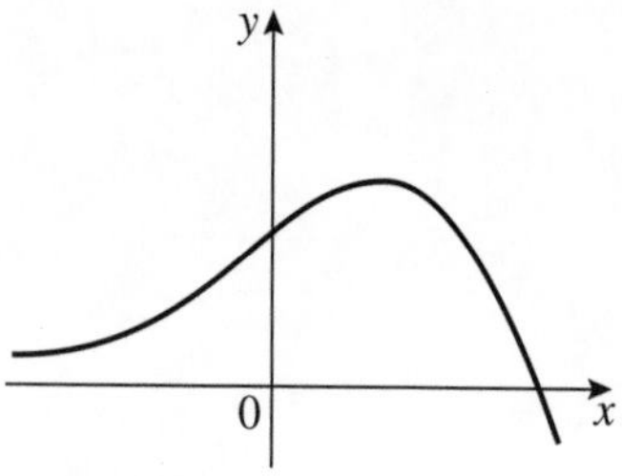

maximum at $(\frac{1}{2}, 2)$
curve crosses the x-axis at $(\frac{3}{2}, 0)$ and y-axis at (0, 1)

Revision exercise 5

1 $x - 5y + 14 = 0$ **2** $y = -4x - 17$ **3** $-\frac{1}{3}$

4 (a) $y = -\frac{1}{3}x + 4$ **(b)** 24

5 (a) $\frac{3}{4}$ **(b)** $(0, -3)$ **(c)** $3x - 4y + 10 = 0$

6 (a) $2x + 3y - 18 = 0$ **(b)** $(9, 0)$ **(c)** $3x - 2y - 27 = 0$

7 (a) $\frac{3}{2} \times -\frac{3}{2} \neq -1$ **(b)** $(12, 0), (6, 9)$ **(c)** 54

8 (a) $-\frac{1}{2}$ **(b)** $y = 2x - 8$ **(c)** $(4, 0)$

9 (a) $y = 2x - 4$ **(b)** $(\frac{8}{5}, -\frac{4}{5})$ **10 (a)** $4x + 5y + 26 = 0$

11 **(a)** $y = \frac{5}{6}x - \frac{13}{6}$ **(b)** -3

12 **(a)** **(i)** 22 **(ii)** 2 **(b)** $y = x + 6$

13 **(a)** $-\frac{1}{3}$ **(b)** 5 **(c)** $y = -x + 3$

14 **(a)** $\frac{1}{2}$ **(b)** $y = \dfrac{1}{2}x + \left(\dfrac{3k}{2} - 3\right)$

15 **(a)** $-4 \times \frac{1}{4} = -1$ **(b)**

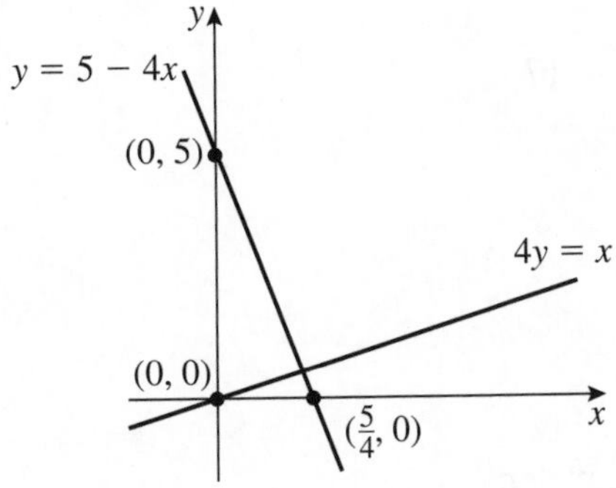

15 **(c)** $\left(\frac{20}{17}, \frac{5}{17}\right)$

Revision exercise 6

1 $-1, 2, 5$ **2** $n = 13$

3 **(a)** $2k - 12$ **(b)** $2k^2 - 12k - 6$ **(c)** $k = 7, -1$

4 **(a)** 55 **(b)** -1 **5** **(a)** 1050 **(b)** 1045 **6** $n = 28$

7 1200, 3750 **8** $a = 17.5, d = -0.5$

9 **(a)** $a = 4$ **(b)** $S_{10} = -5$

10 **(a)** $d = 4$ **(b)** $a = 10$

11 **(b)** 80 200

12 **(a)** £2540 **(b)** £50 100 **(c)** 6 years

13 **(a)** £2450 **(b)** £59 000 **(c)** 30

14 **(b)** 3700 **(c)** £91 000

Revision exercise 7

1 $3x^2 - 14x + 15$ **2** $6x - \frac{5}{2}x^{-\frac{1}{2}} - x^{-3}$ **3** $8x^3 + \frac{3}{2}x^{\frac{1}{2}} + x^{-\frac{1}{2}}$

4 14 and -16 **5** -5, 11 and 16 **6** $(2, -2)$ and $(-2, 6)$

7 $\left(\frac{1}{3}, \frac{4}{27}\right)$ and $(-3, 12)$ **8** $12 - 12t, -12$

9 **(a)** $2\pi r - \dfrac{729}{r^2}$ **(b)** $r = \sqrt[3]{\dfrac{729}{2\pi}}$ **10** 10, 6280

11 $a = -5, b = 2$

12 **(a)** $A = \frac{3}{2}, n = 2, B = 16$ **(b)** $(4, 32)$ **(c)** $2y + 45x = 143$

13 **(a)** $x - \dfrac{16}{x^3}$ **(b)** $x = \pm 2$

14 **(b)** $1 + 6x^{-2} - 27x^{-4}$ **(c)** $x = \pm\sqrt{3}$

15 **(a)** $x(x - 1)(x - 5)$ **(b)** $1, 5$ **(c)** -4

16 **(b)** $y + 9x = 54$

17 **(a)** $9, -6$ and 1 **(b)** 4 **(c)** $4y + x = 17$

18 **(a)** $6x - 4 + \frac{2}{3}x^{-2}$ **(b)** $5y = 3x + 7$

Revision exercise 8

1 $\frac{1}{2}x^3 - \frac{1}{16}x^4 + c$ **2** $-x^{-2} - 2x^{\frac{5}{2}} + c$

3 $x^2 - 2x^{\frac{1}{2}} + c$ **4** **(a)** $-x^{-1} - 2x + \frac{1}{3}x^3 + c$

5 $x^3 - \frac{5}{2}x^2 - 2x + c$

6 **(a)** $\frac{5}{2}x^2 + \frac{4}{3}x^{\frac{3}{2}} + c$ **(b)** $\frac{25}{3}x^3 + 8x^{\frac{5}{2}} + 2x^2 + c$

7 **(a)** $P = 2, Q = 1, R = -3$ **(b)** $\frac{4}{5}x^{\frac{5}{2}} + \frac{2}{3}x^{\frac{3}{2}} - 6x^{\frac{1}{2}} + c$

8 **(a)** $P = 2, Q = 3, R = 1$ **(b)** $\frac{4}{5}x^{\frac{5}{2}} + 2x^{\frac{3}{2}} + 2x^{\frac{1}{2}} + c$

9 $2x^3 + 4x^{-1} + \frac{3}{4}x^{\frac{4}{3}} + c$ **10** $3x^4 - 2x^{\frac{3}{2}} + 5x + c$

11 $\frac{3}{5}x^{\frac{5}{3}} - 2x + 3x^{\frac{1}{3}} + c$ **12** $2x^{\frac{3}{2}} - 2x^{\frac{1}{2}} - 5x + 11$

13 $\dfrac{x^4}{4} + x^2 - 6x + 9$

14 **(a)** $x^3 - x^2 - 6x$ **(b)**

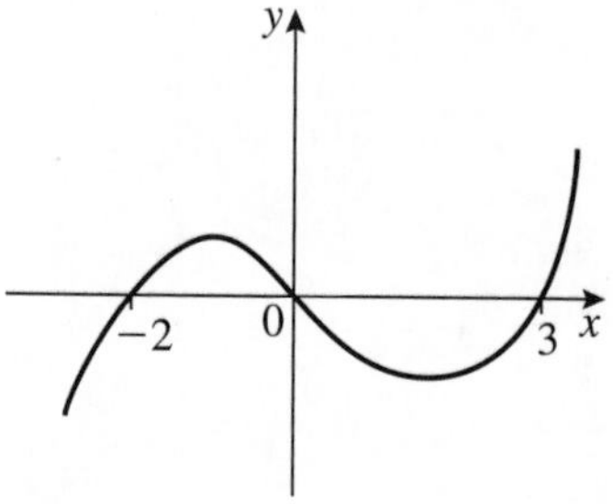

15 **(a)** $x^3 - 2x^2 + x$ **(b)**

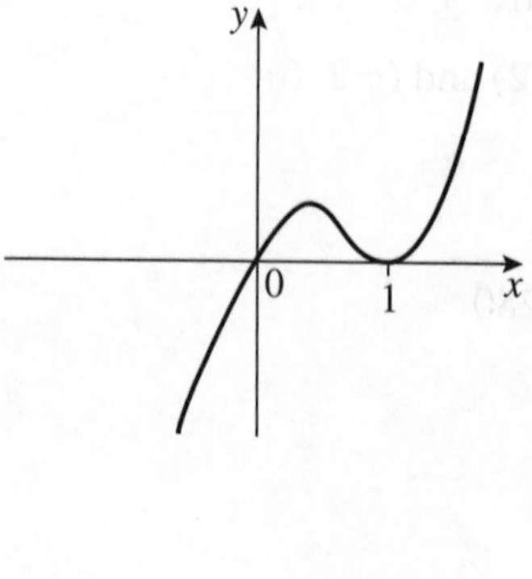

Examination style paper

1 $x > -3\frac{1}{2}$ **2 (a)** 8   **(b)** 28

3 (a)

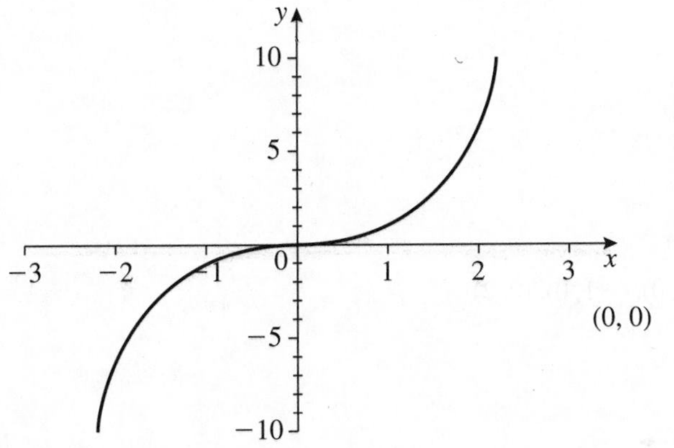

(0, 0)

3 (b)

(−1, 0), (0, 1)

3 (c)

(2, 0), (0, −8)

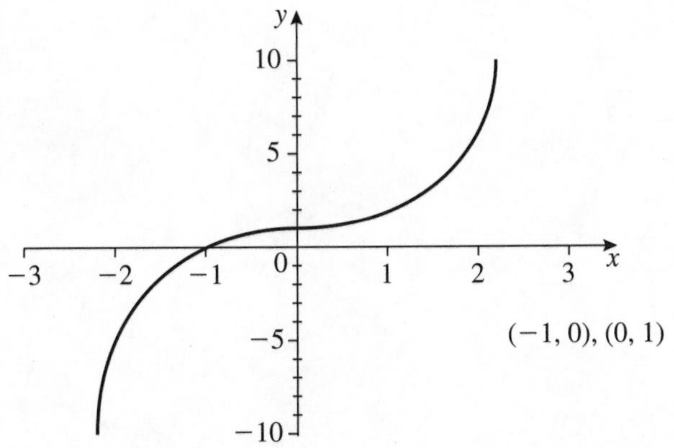

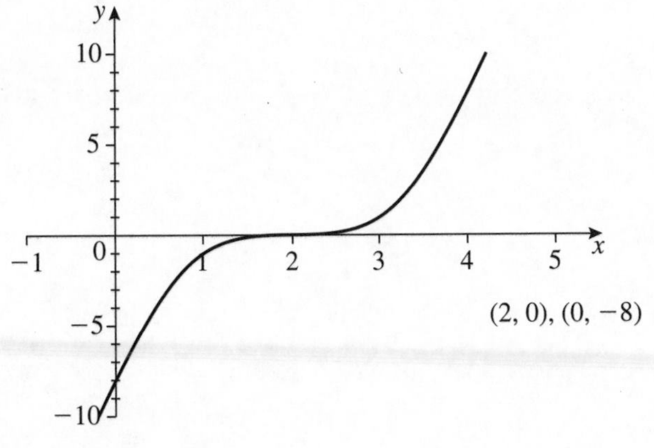

4 (a) $13 + 4\sqrt{3}$ **(b)** $2 + \sqrt{3}$

5 (a) $12x^2 + 3x^{-4}$ **(b)** $x^4 + \dfrac{1}{2x^2} + C$

6 $f(x) = 9x + \dfrac{x^2}{2} - 4x^{\frac{3}{2}} - 5$

7 (a) $a = 51, d = -\frac{3}{2}$ **(b)** 366   **(c)** 69

8 (a) $6x^2 - 2 - 4x^{-2}, 12x + 8x^{-3}$ **(c)** $y = 21x - 28$

9 (a) $5x + 2y - 17 = 0$

9 (b)

$(-2, 0), (-1, 0), (0, 2)$

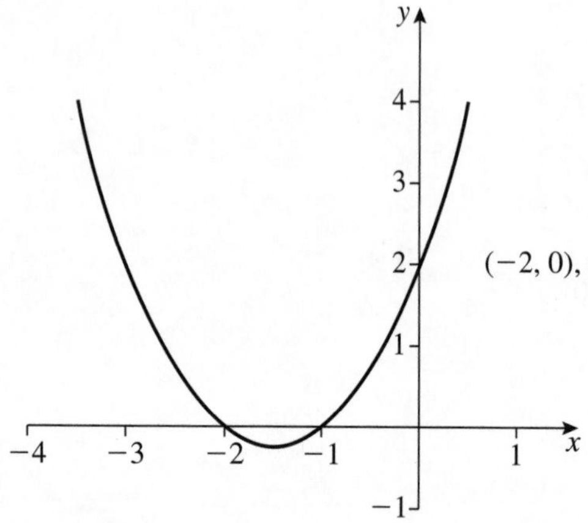

9 (c) $(-\frac{13}{2}, \frac{99}{4})$ and $(1, 6)$

10 (a) $-2, 6$ **(b)** $k = -2: x = 1$ $k = 6: x = -3$

10 (c) $(x + 4)^2 - 5$ **(d)** $-4 \pm \sqrt{5}$